Contents

Preface

Biology for Edexcel International GCSE Workbook is the new edition of the Edexcel International GCSE Biology Practice Book. It is designed as a 'write-in' workbook for students to practise and test their knowledge and understanding of the content of the International GCSE Biology course.

The sections are presented with the same headings and in the same order as in the Student Book, *Edexcel International GCSE Biology Student Book Second Edition*.

The Workbook should be used as an additional resource throughout the course alongside the Student Book. The 'write-in' design is ideal for use by students in class or for homework.

Answers can be found online at www.hoddereducation.co.uk/igcsebiology

Extended writing questions

- The mark allocation for extended writing questions is usually 6 marks. This gives a guide as to how long you should spend on this question in relation to the rest of the paper – probably about 6 minutes.

- Before starting to write, pick out any key words in the question. Mark them in some way and think carefully about what is being asked.

- Use your time sensibly by writing down a brief plan or a list of what you want to include. You can afford to spend 3 minutes on planning and deciding what you are going to write, as 3 minutes will be enough to write an answer that gains full marks. Don't spend too long or you could run out of time for other questions.

- When answering this type of question it is preferable to make sure that you give a complete written description rather than relying entirely on a diagram. You can provide explanations in a written account that are more difficult to convey on a diagram. On the other hand, a diagram can help you to remember the key points, and a clear diagram with appropriate annotations can gain high marks.

- Skills being tested include the ability to select and organise information. Facts must be correct and relevant to the question.

- Often a single word is not enough for a mark but requires further description or elaboration in some way. The question may expect you to make links between different aspects of a topic, so make sure those links are clear in what you write.

- The question may refer to two different parts of a topic – say, features or processes to compare. Make sure you cover both aspects, otherwise you will not be able to gain full marks.

- No marks are awarded for the quality of written work (such as spelling, grammar and organisation of subject matter), but you are more likely to include all the relevant material if you organise it sensibly and logically. It is also easier for the examiner to read if your handwriting is legible and written in sentences that are clearly set out.

International GCSE (9–1)

ERICA LARKCOM
ROGER DELPECH
KATHY EVANS

Biology

for Edexcel International GCSE

WORKBOOK

HODDER EDUCATION
LEARN MORE

Orders: please contact Hachette UK Distribution, Hely Hutchinson Centre, Milton Road, Didcot, Oxfordshire, OX11 7HH. Telephone: +44 (0)1235 827827. Email education@hachette.co.uk Lines are open from 9 a.m. to 5 p.m., Monday to Friday. You can also order through our website: www.hoddereducation.co.uk

Photo credit: **p.31** © Roger Delpech.

© Erica Larkcom, Roger Delpech and Kathy Evans 2017

First published in 2017 by
Hodder Education
An Hachette UK Company
Carmelite House
50 Victoria Embankment
London EC4Y 0DZ

www.hoddereducation.co.uk

Impression number 10 9

Year 2022

Illustrations by Aptara

Typeset in Frutiger 55 roman 10/13 pt by Integra Software Services Pvt. Ltd., Pondicherry, India

Printed in the UK

A catalogue record for this title is available from the British Library.

ISBN: 978 1 5104 0565 3

Experimental design questions

These are the 'experimental design' questions, often known as the CORMS questions. If we look a little further at this name, we see how it helps provide a framework for your answer. The letters are given in the mark scheme and it helps you to use them in your answers. In this way, you can check that you have included reference to all the necessary factors that should be considered in designing the investigation or experiment asked for in the question.

We now look at each letter in turn.

C	= what is being **C**hanged (or **C**ompared) in the experiment. This is the independent variable and also covers the idea of a **C**ontrol. In the question on bananas ripening, on page 91, the change is the presence (or not) of ethene from the ripe tomato.
O	= the **O**rganism being used and some statement about it to make sure the investigation is valid. Often this is covered by reference to using the same species or variety so that the effect of the change can be judged fairly. So, in the question on bananas ripening, all bananas should be of the same variety or from the same batch and starting from the same stage of ripeness.
R	= **R**eplication or repeats, so that several results are obtained rather than relying on a single measurement or observation. This is good practice in any experimental work.
M1+**M2**	= the **M**easurements taken. This is the dependent variable, because it 'depends' on what the change is when setting up the experiment. Often this may refer to a change in mass, or height or something you can measure in numbers. For the question on bananas, you decide on an 'end-point' on the colour scale (which represents the stage of ripeness). A second **M** mark is usually given for reference to a timescale – in the banana question, this is how long it takes the banana to reach the colour chosen for the end-point. It is important to suggest an actual time – this shows you are thinking about whether the change occurs in seconds, or hours or perhaps weeks. You may not know the correct time, but make a sensible attempt. In the other questions on page 91, the cabbages treated with fertiliser are likely to take several weeks before reaching a stage to be weighed, whereas the yoghurt investigation is likely to be completed in a few hours.
S1+**S2**	= a variable that must be kept the **S**ame or controlled in this experiment. Such factors may include the quantity used (same volume, same mass) or other variables such as same temperature, same humidity or whatever is appropriate for the investigation. Usually there are plenty of factors you could choose, but make sure it is relevant to the investigation.

Lastly, the topic in the question may be a novel one, so perhaps not something you have already studied in your specification, but you should be able to apply and adapt these principles to any of the questions set. Spend a couple of minutes thinking how you would plan the investigation, and then try to follow through the CORMS letters to make sure you cover all essential aspects of the design.

Living organisms: variety and common features

1.1: Using and interpreting data

1   A student was asked to find out whether differences in pH have an effect on the activity of an enzyme that digests protein. The enzyme used was one that is found in the stomach of mammals.

Seven test tubes, each containing a protein solution plus 2% solutions of the enzyme, were placed in a water bath kept at 37 °C. The solutions covered a range of pH from 3 to 9. At the end of the experiment, the student calculated the rate at which a standard amount of protein had been digested in each tube.

pH	Rate of protein breakdown / arbitrary units
3	100
4	96
5	81
6	60
7	5
8	0
9	0

The results are shown in the table.

a) Plot a graph of these results on a graph grid 11 cm × 11 cm. Join the points with straight lines.
   (Use a separate sheet of graph paper to plot your graph.)   *[4 marks]*

b) Use the graph to estimate the pH at which the rate of protein breakdown would have been 70 arbitrary units.   *[1 mark]*

....

c) i) At which pH does the enzyme used in the experiment work best?   *[1 mark]*

....

   ii) Explain, using the term **active site**, why the activity of the enzyme changes with changes in pH.   *[3 marks]*

....

....

....

....

d) Suggest how the results for pH 4 would be different if the experiment was carried out at 20 °C. Explain your answer.   *[3 marks]*

....

....

....

....

e) Describe how you would test the original solution for the presence of protein.　　*[2 marks]*

....

....

....

(Total = 14 marks)

2　A student investigated the effect of different concentrations of salt solution on the mass of potato cylinders. He cut potato cylinders from a fresh potato using a cork borer. He blotted each cylinder to remove excess water and weighed it.

He then put each cylinder into a beaker containing salt (sodium chloride) solution for one hour. The student removed the cylinders, blotted them and re-weighed them. For each cylinder, he noted the change in mass and converted it to a percentage. His results are shown in the table.

Concentration of salt solution / M	Initial mass / g	Final mass / g	Change in mass / g	Percentage change in mass
0.0 (water)	2.1	2.6	+0.5	+24
0.1	1.9	2.2	+0.3	
0.2	2.0	2.1		+5
0.4	2.2	1.8		−18
0.6	2.0	1.4	−0.6	−30
0.8	1.9	1.1	−0.8	−42

Note: Molarity (moles, M) is a measure of concentration.

a) Some of the data are missing from the table.

　i)　Calculate the changes in mass of the potato cylinders in 0.2 M and 0.4 M salt solution. (Write your answers in the spaces given in the table.)　　*[2 marks]*

　ii)　Calculate the percentage change in mass for the potato cylinder in 0.1 M salt solution. (Write your answer in the space given in the table.)　　*[2 marks]*

b) The student drew a line graph of the results.

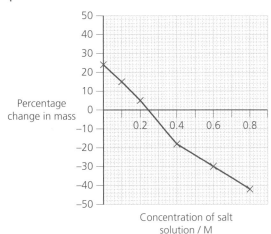

Use information from the graph to answer the following questions.

　i)　What would be the expected percentage change in mass of the potato cylinder in 0.5 M salt solution?　　*[1 mark]*

....

ii) In what concentration of salt solution would potato cylinders neither gain nor lose mass? *[1 mark]*

....

iii) Over what range of salt concentrations did the potato cylinders lose mass? *[1 mark]*

....

c) Use your knowledge of osmosis to explain why potato cylinders in these solutions lost mass.

[3 marks]

....

....

....

....

....

(Total = 10 marks)

1.2: Practical activities

1   A piece of food material may contain carbohydrate, lipid (fat) and protein.

a) Describe how you would prepare a sample of solid food so that you can test it to find out what it contains. *[2 marks]*

....

....

b) The table summarises tests for different food substances. Complete the table to show details of the tests you could use to find out whether the food material contains lipid (fat), glucose, protein or starch. One row has been done for you. *[6 marks]*

Substance being tested for in sample	Reagent to be added	Requires heating (YES / NO)	Appearance / Colour at start	Appearance / Colour of positive test result
lipid (fat)	ethanol	NO	clear solution	cloudy solution
glucose				
protein				
starch				

(Total = 6 marks)

2   A student cut two cubes, A and B, from a block of colourless jelly. The sides of cube A were 1 cm and the sides of cube B were 3 cm.

The cubes were used as models of living organisms. The student submerged each cube in a red dye for 10 minutes. He then removed it and cut it in half. The diagram shows how far the dye had diffused into the jelly cubes.

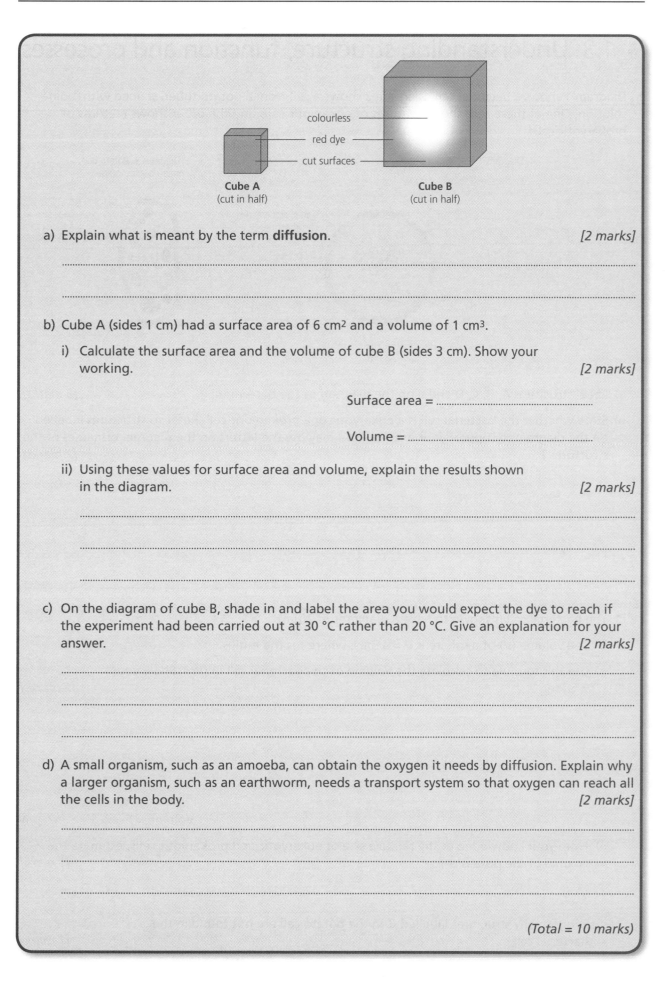

Cube A
(cut in half)

Cube B
(cut in half)

a) Explain what is meant by the term **diffusion**. *[2 marks]*

....

....

b) Cube A (sides 1 cm) had a surface area of 6 cm² and a volume of 1 cm³.

   i) Calculate the surface area and the volume of cube B (sides 3 cm). Show your
      working. *[2 marks]*

                        Surface area = ...

                        Volume = ...

   ii) Using these values for surface area and volume, explain the results shown
       in the diagram. *[2 marks]*

....

....

....

c) On the diagram of cube B, shade in and label the area you would expect the dye to reach if
   the experiment had been carried out at 30 °C rather than 20 °C. Give an explanation for your
   answer. *[2 marks]*

....

....

....

d) A small organism, such as an amoeba, can obtain the oxygen it needs by diffusion. Explain why
   a larger organism, such as an earthworm, needs a transport system so that oxygen can reach all
   the cells in the body. *[2 marks]*

....

....

....

(Total = 10 marks)

1.3: Understanding structure, function and processes

1 Diagram 1 shows a bacterial cell. Diagram 2 shows a cell from a potato tuber, stained with iodine solution. One of these cells shows features of eukaryotic cells and the other shows features of prokaryotic cells.

Diagram 1: Bacterium

Diagram 2: Potato cell stained with iodine

C.....................................

single circular chromosome

B.....................................

D.............................

A.....................................

E.............................

F

G

2 μm

a) Label structures A, B, C, D and E on the diagram of the bacterium.   *[5 marks]*

b) State whether the bacterial cell is a eukaryotic or a prokaryotic cell. Refer to structures labelled on the diagram to support your answer. (You may use the letters on the diagram or names of the structures.)   *[2 marks]*

....

....

....

....

c) The bacterial cell is shown to have a diameter of 2(μm).

i) The volume (V) of a sphere is $V = 4/3\pi r^3$, where r is the radius.

Use this formula to calculate the volume of the bacterial cell in μm^3. Show your working.   *[2 marks]*

Volume = ... μm^3

ii) From your knowledge of the relative size of eukaryotic and prokaryotic cells, estimate the volume of the potato cell.   *[1 mark]*

....

iii) Suggest why structures labelled G in the potato cell are not found in the bacterial cell.   *[1 mark]*

....

d) i) The molecule found in the structures labelled A is found in chromosomes and is also present in some viruses. Name this molecule. *[1 mark]*

....

ii) Suggest why viruses must invade a eukaryotic or prokaryotic cell in order to reproduce. *[2 marks]*

....

....

....

(Total = 14 marks)

2 Look at diagram 2 in Question 1 above.

The diagram shows a cell from a potato tuber, stained with iodine solution.

a) List **two** ways in which plant cells differ from animal cells. *[2 marks]*

....

....

b) i) Describe the function of the structures labelled G. *[2 marks]*

....

....

....

ii) Name **one** structure that would be found in a leaf cell in a potato plant but is not shown in the potato cell in this diagram. *[1 mark]*

....

c) The structures labelled F have stained a blue-black colour.

i) Name the chemical substance found in structure F. *[1 mark]*

....

ii) Give **one** function of structure F in the potato plant. *[1 mark]*

....

iii) The chemical substance in F is not usually found in fungi. Name a chemical substance found in fungi that has a similar function to F in potato plants. *[1 mark]*

....

(Total = 8 marks)

3 a) i) Draw and label a typical human cheek cell as seen with the light microscope. *[3 marks]*

ii) List **two** ways in which a yeast cell differs from a human cheek cell. *[2 marks]*

....

....

b) The human body contains more than a hundred different types of cells.

i) Give the term used to describe a group of similar cells. *[1 mark]*

....

ii) Describe how different cells are organised into a system, such as the nervous system, in the body. Give **named** examples of cells to support your answer. *[3 marks]*

....

....

....

....

c) i) Some cells in the body are described as 'stem cells'. What is the role of stem cells in the body? *[2 marks]*

....

....

....

ii) Discuss ways in which stem cells may be of importance in medicine. *[3 marks]*

....

....

....

....

(Total = 14 marks)

1.4: Applying principles

1 a) The diagram shows the structure of a cell from a potato.

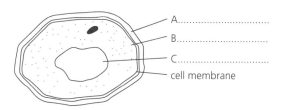

A..............................

B..............................

C..............................

cell membrane

On the diagram, name the parts of the cell labelled A, B and C. The cell membrane has been labelled for you.   *[3 marks]*

b) A student investigated osmosis in potatoes. She cut a large potato into halves and removed about 1 cm of peel from the edge of the cut surface. She boiled one potato half for 10 minutes and then cooled it. She cut a well in the top of each potato half and placed them in dishes of water, as shown in diagram 1. The student put 5 cm³ of concentrated sucrose (sugar) solution inside each well, and left the dishes for four hours.

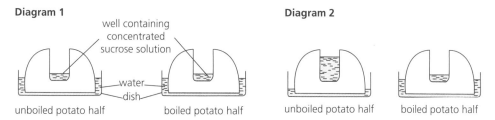

Diagram 1

well containing concentrated sucrose solution

water

dish

unboiled potato half   boiled potato half

Diagram 2

unboiled potato half   boiled potato half

i) Diagram 2 shows the results for the potato halves after four hours. Describe the results for the unboiled potato.   *[2 marks]*

....

....

....

ii) Suggest an explanation for the results of the unboiled potato.   *[3 marks]*

....

....

....

iii) Suggest a reason for the liquid level staying the same in the boiled potato.   *[2 marks]*

....

....

....

(Total = 10 marks)

2 In an investigation, the effect of osmosis on red blood cells was observed. A drop of blood was placed on each of two microscope slides. On slide A, the blood was mixed with a 0.85% solution of salt (sodium chloride). On slide B, the blood was mixed with a 3.0% solution of salt. Both slides were observed under the microscope.

The appearance of one cell from each slide, at the start and again after a few minutes, is shown in the diagram.

slides A and B
red blood cells at the start
of the investigation

slide A
red blood cells
after a few minutes
in 0.85% salt solution

slide B
red blood cells
after a few minutes
in 3.0% salt solution

a) i) Describe the appearance of the red blood cells on slides A and B after a few minutes in the salt solutions.   *[2 marks]*

....

....

....

ii) Suggest an explanation for the changes in the cells on slide B.   *[3 marks]*

....

....

....

....

....

iii) The red blood cells circulating in the plasma do not change shape in this way. Suggest a reason for this.   *[2 marks]*

....

....

....

b) Red blood cells take up some substances that they need by active transport. How does this differ from osmosis?   *[2 marks]*

....

....

....

(Total = 9 marks)

3   A student prepared a starch agar jelly by stirring a solution of starch into hot, liquid agar. It formed a colourless jelly when cooled. She cut the agar jelly into small cubes with sides of 1 cm and large cubes with sides of 2 cm.

She put three cubes of each size into a beaker containing amylase solution. After 30 minutes she removed the cubes and cut each one carefully in half.

She covered the cut surfaces with iodine solution and observed the colour changes. The diagram shows the colours of the cut surfaces after iodine was added.

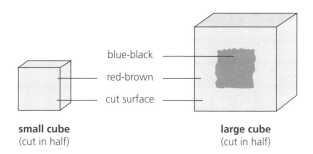

small cube
(cut in half)   **large cube**
                                 (cut in half)

cubes of starch agar jelly after 30 minutes in
amylase solution

a) What does the appearance of the small and large cubes suggest about the distribution of
    starch?   *[2 marks]*

    ....

    ....

    ....

b) Suggest an explanation for the results of the small cube.   *[2 marks]*

    ....

    ....

    ....

c) Suggest an explanation for the difference in starch distribution between the small and
    large cubes.   *[3 marks]*

    ....

    ....

    ....

    ....

    ....

    ....

(Total = 7 marks)

2 Nutrition and respiration

2.1: Using and interpreting data

1 Lipids (fats) in the diet are an important source of energy and some provide essential molecules for cell membranes. Butter, margarine and low-fat spreads are all important sources of fat in people's diets. The graph shows changes in the fats eaten in Britain from 1976 to 1996.

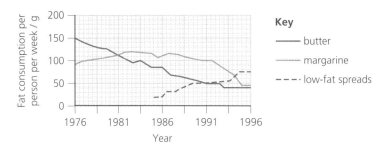

a) i) What was the largest source of fat in the weekly diet in 1976? [1 mark]

....

ii) The total fat eaten per person per week in 1976 was 240 g. How many grams came from margarine? [1 mark]

....

iii) From the graph, describe the change in the amount of butter eaten per week over the 20 years from 1976 to 1996. [2 marks]

....

....

....

b) i) From the graph, calculate the total fat eaten per person per week in 1991. (Show your working.) Give **one** way in which this differed from the total amount of fat eaten per person per week in 1976. [3 marks]

Total fat eaten per person per week = ...

ii) Suggest **two** reasons why the amount of fat in the weekly diet in 1996 is better for health than that in 1976. [2 marks]

....

....

(Total = 9 marks)

2 Students in a class used the apparatus shown to measure the energy content of different foods.

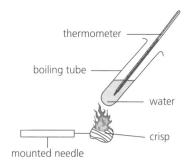

One student investigated the energy value of potato crisps. She put 20 cm³ of water into the boiling tube. She then weighed one of the crisps, set it alight with a Bunsen burner and quickly held it underneath the boiling tube. She recorded the original temperature of the water and the temperature immediately after the crisp finished burning.

The student used the formula below to calculate the energy released when one crisp was burned.

energy in joules = mass of water × rise in temperature × 4.2

Note: 4.2 joules = the energy required to raise the temperature of 1 g of water by 1 °C.
1 cm³ of water has a mass of 1 g.

a) i) The initial temperature of the water was 20 °C and the final temperature was 52 °C. Calculate the energy, in joules, released when the crisp was burned. Show your working.
[3 marks]

Energy = ... joules

ii) The mass of the crisp was 0.8 g. Calculate the energy released per g of crisps. Show your working.
[1 mark]

Energy = ... joules

b) i) The results obtained by the students for other foods are shown in the table.

Food	Energy per g / joules
rice cakes	1600
crispbread	1450
raisins	1300

Calculate the difference in energy released when 1 g of rice cake is burned compared with 1 g of raisins. Show your working.
[2 marks]

Difference in energy = ... joules

ii) The students saw that the energy value of the rice cakes was given on the packaging as 1582 kJ per 100 g. Suggest a reason for the difference between the results obtained by the students and the actual value. *[2 marks]*

....

....

....

(Total = 8 marks)

3 The carbohydrate, lipid (fat) and protein content in 10 g of two foods is shown in the table.

Food	Mass per 10 g portion / g		
	Carbohydrate	Lipid (fat)	Protein
white bread	4.8	0.2	0.7
eggs (boiled)	0.0	1.1	1.3

a) A slice of bread weighs 40 g. An athlete in training is advised to include about 116 g of carbohydrate per meal. How many slices of bread would provide this? Show your working. *[3 marks]*

Number of slices of bread = ...

b) The athlete was advised to include about 77 g of protein in his diet every day.

i) Explain why an athlete needs protein. *[2 marks]*

....

....

....

ii) What is the percentage of protein in white bread? Show your working. *[2 marks]*

Percentage of protein = ...

iii) Suggest why it would be preferable for the athlete to obtain the protein he needs from foods other than bread. Give reasons to support your answer. *[2 marks]*

....

....

....

(Total = 9 marks)

2.2: Practical activities

1   A student investigated the factors needed for a plant to carry out photosynthesis. He used a plant that had been kept in the dark for two days. He covered some of the leaves with bags, as follows:

Leaf 1: Transparent polythene bag containing air

Leaf 2: Transparent polythene bag containing air and a substance that absorbs carbon dioxide

Leaf 3: Black polythene bag containing air

a)   Why does a plant become 'destarched' when left in the dark for two days?   *[1 mark]*

....

b)   Name a substance that could be used to absorb carbon dioxide.   *[1 mark]*

....

c)   The student left the plant in the light for eight hours, and then took the leaves off the plant and tested them for starch. The steps he used are shown in the diagram.

Step 1

| Leaf dipped in boiling water |

Step 2

| Leaf boiled in ethanol |

Step 3

| Leaf dipped in hot water |

Step 4

| Leaf spread on a white tile and covered with iodine solution |

   i)   In step 1, why is the leaf dipped in boiling water? *[1 mark]*

....

....

   ii)   What is the purpose of step 2?   *[1 mark]*

....

....

   iii)   Describe **two** precautions that should be taken when carrying out step 2.   *[2 marks]*

....

....

....

....

   iv)   In step 4, what colour would the leaf be if starch was present, and if starch was absent?   *[2 marks]*

....

....

....

....

d) The table shows the results of the starch tests.

	Leaf 1	Leaf 2	Leaf 3
Starch present	✓	✗	✗

Write a suitable conclusion for these results. *[2 marks]*

....

....

....

e) If the experiment had been carried out using a plant with variegated leaves, what differences would you expect for the presence of starch in the green and white regions of the variegated leaves? Explain your expected results. *[3 marks]*

....

....

....

....

....

....

(Total = 13 marks)

2 A student investigated the rate of photosynthesis at different temperatures. For experiment 1, she set up the apparatus shown in the diagram and counted the number of bubbles of oxygen produced per minute at each temperature. A lamp placed near the apparatus provided light.

For experiment 2, the student moved the lamp 10 cm closer to the beaker and repeated the readings.

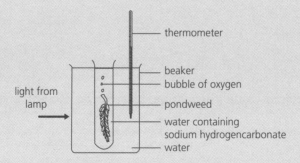

a) Why did the student add sodium hydrogencarbonate to the water in the test tube? *[1 mark]*

....

b) The student took five readings at each temperature and recorded the mean number of bubbles per minute in a table.

Temperature / °C	Number of bubbles per minute	
	Experiment 1	Experiment 2
5	15.0	15.2
10	21.2	21.8
15	27.0	27.6
20	29.0	34.0
25	28.6	38.2
30	29.0	42.0

i) Plot a graph of the data (on a graph grid 9 cm × 9 cm) using straight lines to join the points. (Use a separate sheet of graph paper to plot your graph.) [6 marks]

ii) From your graph, what is the rate of photosynthesis at 18 °C (in number of bubbles per minute) for experiment 1 and experiment 2? [2 marks]

....

....

c) i) Describe the differences between the two graphs for temperatures from 20 °C to 30 °C. [2 marks]

....

....

....

ii) Suggest an explanation for the differences. [2 marks]

....

....

....

d) Suggest **one** precaution that the student could take to ensure that the results are as accurate as possible. [1 mark]

....

....

e) Describe how you could modify this experiment to investigate the effect of different carbon dioxide concentrations on the rate of photosynthesis. [2 marks]

....

....

....

....

(Total = 16 marks)

3 A student set up the apparatus shown to measure the rate of respiration of living organisms at different temperatures. She did two experiments, the first with maggots and the second with peas. The respiration rate can be measured by recording the rate at which the coloured water rises in the glass tube.

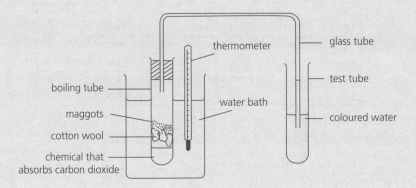

Experiment 1: Sara put eight maggots into the boiling tube. She made sure that the level of the coloured water in the glass tube was level with the water outside the tube, at the start. After 30 minutes, she measured the height that the water had risen to. She repeated this procedure at different temperatures and recorded the results.

Experiment 2: Sara repeated the experiment, replacing the maggots with an equivalent mass of peas that had been soaked in water overnight.

a) i) Name a chemical that absorbs carbon dioxide.   *[1 mark]*

....

   ii) Explain why the level of coloured water in the glass tube rises.   *[2 marks]*

....

....

....

b) Sara's results for both experiments are shown in the table.

Temperature / °C	Height of water / mm	
	Experiment 1 (maggots)	Experiment 2 (peas)
5	8	3
10	12	5
15	17	7
20	23	11
25	31	17

The results for the maggots have been plotted in the graph. On the same graph, plot the results Sara obtained for the peas, joining the points with straight lines.   *[2 marks]*

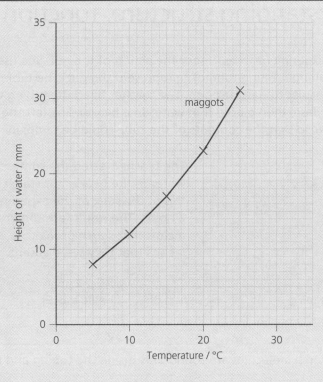

c) i) What was the rate of respiration of the maggots (in mm per minute) at 20 °C?   *[2 marks]*

....

....

ii) From the graph of the results for maggots, suggest what you would expect the water height to be at 30 °C.   *[1 mark]*

....

d) i) How much faster did the maggots respire at 18 °C compared with 8 °C? Show your working.   *[3 marks]*

Answer = ...

ii) Give an explanation for this based on your knowledge of enzyme action.   *[2 marks]*

....

....

....

(Total = 13 marks)

● 2.3: Understanding structure, function and processes

1 A teacher set up an experiment to represent the action of the gut. She used a length of partially permeable dialysis tubing (Visking tubing) to represent the gut. The tubing has small pores that allow only small molecules to pass through. The teacher filled the tubing with starch solution mixed with the enzyme amylase, then placed it in water in a beaker. She left the open end hanging over the side of the beaker so that the contents of the tube could be sampled.

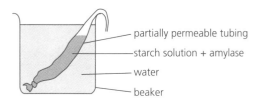

partially permeable tubing

starch solution + amylase

water

beaker

The teacher tested the contents of the tubing and the water in the beaker for starch and for maltose. She did this at the start of the experiment and again after 2 hours. The results are shown in the table. A tick (✓) means that starch or maltose was present and a cross (✗) means that they were absent.

Test	At start		After 2 hours	
	Inside tubing	Water in beaker	Inside tubing	Water in beaker
starch	✓	✗	✗	✗
maltose	✗	✗	✓	✓

a) i) Suggest why starch molecules are found only inside the tubing and not in the water in the beaker. *[1 mark]*

....

 ii) Explain why, after 2 hours, maltose is present inside the tubing. *[2 marks]*

....

....

....

....

 iii) Explain why, after 2 hours, maltose is present in the water in the beaker. *[2 marks]*

....

....

....

b) i) Describe **two** ways in which this experiment represents the action of the human gut. *[2 marks]*

....

....

....

....

ii) Describe **two** ways in which the wall of the small intestine is better adapted than the Visking tubing for encouraging efficient movement of digested substances across it.   *[2 marks]*

....

....

....

....

(Total = 9 marks)

2   The diagram shows a section through a leaf.

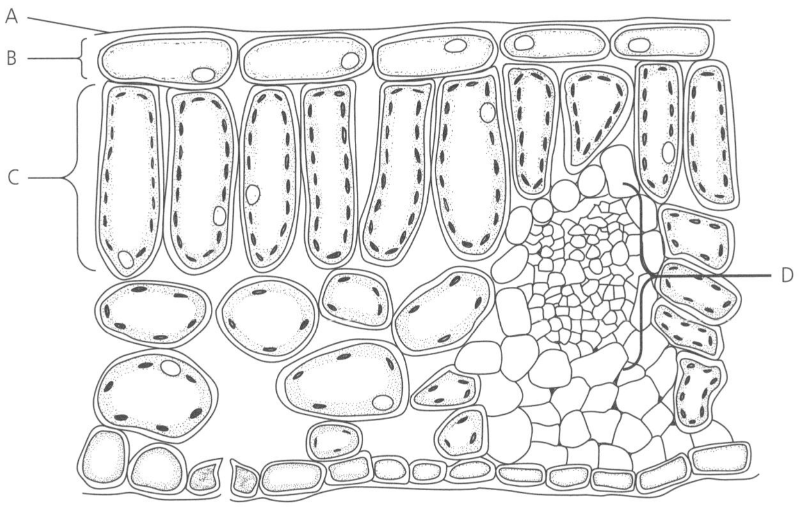

a) Complete the table by naming parts A, B, C and D and giving a function for each.   *[4 marks]*

Letter	Name of part	Function of part
A		
B		
C		
D		

b) i)   On which surface of this leaf are the stomata found?   *[1 mark]*

....

ii) Draw arrows on the diagram to show the pathway taken by a molecule of carbon dioxide during the day, as it passes from the atmosphere to a chloroplast in a palisade mesophyll cell.
     [2 marks]

iii) Explain why this route would change at night.   *[2 marks]*

....

....

....

....

(Total = 9 marks)

3 A student set up the apparatus shown and observed it over 2 hours.

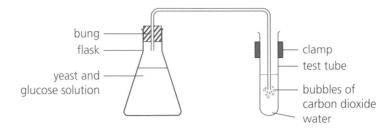

At the beginning, oxygen was available, in the air space and dissolved in the solution. Bubbles of carbon dioxide were produced at a rapid rate.

After 2 hours, there was no oxygen available in the flask. Bubbles of carbon dioxide were observed, but they were produced at a much slower rate.

a) Complete the word equations below to show the processes taking place after 1 hour and after 2 hours, and name each process.

   i) After 1 hour   *[3 marks]*

       .... + oxygen →........................... + ... + energy

       Name of process: ....

   ii) After 2 hours   *[3 marks]*

       .... → ....+ .... + energy

       Name of process: ....

b) Both processes release energy. How would the energy released per minute after 1 hour compare with the energy released per minute after 2 hours?   *[1 mark]*

   ....

                                                                              (Total = 7 marks)

● 2.4: Extended writing

Write your answers to the extended writing questions on separate sheets of paper. First read the general advice on page 4.

1 Describe how the small intestine is adapted for the absorption of small food molecules produced by digestion.   *[6 marks]*

2 Explain how a leaf of a flowering plant is adapted for photosynthesis.   *[6 marks]*

3 Describe how a molecule of starch, taken into the mouth in food, becomes a molecule of glycogen in the liver of a human.   *[6 marks]*

3 Movement of substances in living organisms

3.1: Using and interpreting data

1 Algal balls can be made by trapping thousands of single-celled algae in a jelly-like substance. Drops of the jelly mixture are released from a syringe into a liquid to form spherical balls.

A group of students investigated the effect of chlorophyll concentration on the rate of photosynthesis at different light intensities. They used algal balls that had been made containing different numbers of algae, so they contained different amounts of chlorophyll.

The students prepared five specimen tubes containing high chlorophyll algal balls and five tubes containing the same number of low chlorophyll algal balls. Each tube contained the same volume of hydrogencarbonate indicator.

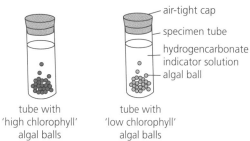

tube with 'high chlorophyll' algal balls

tube with 'low chlorophyll' algal balls

They placed pairs of tubes, as shown in the diagram, in five different light intensities. After 60 minutes, they used a colorimeter to record the colour of the indicator in each tube. The colour was then converted into a measure of the concentration of carbon dioxide in the solution. This was used as a measure of the rate of photosynthesis. Their results are shown in the graph.

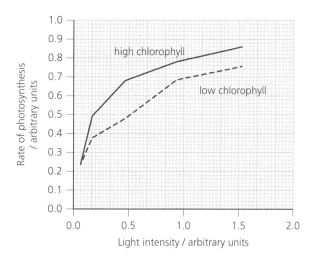

a) i) Write a word equation to show that carbon dioxide is used in photosynthesis. *[2 marks]*

....

ii) Explain why the colour of the hydrogencarbonate indicator after 60 minutes can be used to measure the rate of photosynthesis in the tube. *[3 marks]*

....

....

....

....

b) i) Describe how the rate of photosynthesis for the algal balls with high chlorophyll varies with light intensity. *[3 marks]*

....

....

....

....

....

....

ii) Give **two** differences between the rate of photosynthesis in the high chlorophyll algal balls and the low chlorophyll algal balls. *[2 marks]*

....

....

....

(Total = 10 marks)

2 Many organisations produce information that describes the harmful effects of cigarette smoke and the benefits of stopping smoking. The list below has been extracted from a fact sheet published by an organisation that aims to discourage people from smoking. The information states, for example, that smoking increases the risk of:

- *heart disease, strokes and heart attacks*
- *lung cancer and cancer in other parts of the body*
- *diseases of the chest cavity, including bronchitis and emphysema.*

Graph 1 shows the world cigarette production during the years 1950 to 2000.

Graph 2 shows the cigarette consumption per person in the USA (population about 300 million) during part of the same period (1965 to 2000).

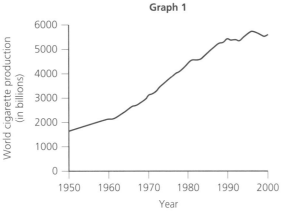
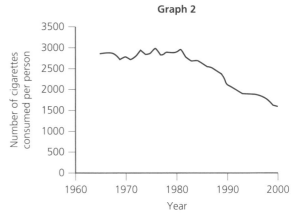

a) i) Suggest **two** reasons for the decrease in cigarette consumption shown in graph 2. *[2 marks]*

....

....

ii) Suggest why, in graph 1, the world production of cigarettes increased from 1950 to 2000 even though, in graph 2, the USA showed a decrease in consumption of cigarettes per person during part of this time. *[2 marks]*

....

....

....

b) Bronchitis is an infection of the lungs and is made worse by mucus that collects in the bronchi. Explain why cigarette smoke may lead to an increase in mucus in the bronchi. *[3 marks]*

....

....

....

....

....

c) Carbon monoxide is contained in cigarette smoke. Describe and explain the effect this may have on a person who smokes. *[2 marks]*

....

....

....

d) Emphysema is a condition of the lungs in which the alveoli become less elastic and their walls start to break down. Explain why a person with emphysema easily becomes breathless. *[2 marks]*

....

....

e) Various lifestyle factors, including smoking, are known to increase the risk of developing coronary heart disease.

Name **two** lifestyle factors, other than smoking, that may influence the development of coronary heart disease. For each factor, state how a person can reduce the chance of premature heart disease. *[2 marks]*

....

....

....

....

(Total = 13 marks)

3 Two students decided to compare their fitness levels. To do this, they compared how quickly their breathing rates recovered after running at a steady speed around their school playing field.
They measured their breathing rates beforehand and then at the end of each minute for 5 minutes immediately following their run. They produced the data shown in the table.

	Breathing rate / breaths per minute					
	Before exercise	After exercise				
		Minute 1	Minute 2	Minute 3	Minute 4	Minute 5
Student A	13	32	27	25	21	16
Student B	14	38	20	16	11	14

a) i) Suggest a suitable hypothesis about fitness levels and post-exercise breathing rate recovery times. *[1 mark]*

....

....

ii) Explain the biological basis for your hypothesis. *[2 marks]*

....

....

....

....

b) Suggest **two** factors the student should have controlled to make sure the experiment is fair. *[2 marks]*

....

....

....

....

c) i) Describe the patterns you see in the data for the students. *[2 marks]*

....

....

....

ii) Suggest which student is fitter and give a reason for your answer. *[1 mark]*

....

....

d) The less fit student decided to follow a six-week period of athletic training. Describe possible changes that the training may make to the body of the student that would contribute to improved fitness. *[4 marks]*

....

....

....

....

....

....

(Total = 12 marks)

3.2: Practical activities

1 Some students investigated the distribution of stomata on leaves of ivy. They used a 'variegated' ivy leaf, some parts of which are green and other parts are yellow. The yellow parts contain no chlorophyll.

The students painted clear nail varnish on the lower surface of the leaf and let it dry. They then pressed on some transparent sticky tape and peeled this off along with the nail polish. This took an impression of the stomata on the surface of the leaf. They placed the peel on a microscope slide and observed it with a microscope.

They decided to make a photomicrograph so they could count the number of stomata in the green and yellow parts and make a comparison.

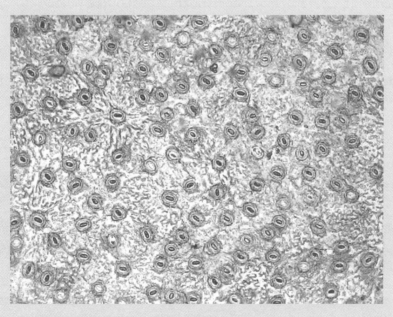

a) i) One student suggested they divide the photomicrograph into four equal sections to make the counting easier. Draw a suitable table that they could use to record their counts in each of the four sections. *[2 marks]*

ii) Divide the photomicrograph into four sections, as suggested by the student, and count the number of stomata in each section. Write your results in the table you have drawn. *[3 marks]*

iii) Describe **two** steps you took while doing the counting to make sure the counts were accurate. *[2 marks]*

....

....

....

....

b) The students then made a similar count of the number of stomata in the peel from a yellow part of the leaf. The two photomicrographs represented the same area of leaf (6.5 mm × 5.0 mm). The total number of stomata that the students counted from this area on a yellow part of the leaf was 101 stomata.

i) Which part of the leaf had more stomata, the green or the yellow? *[1 mark]*

....

ii) Another student was not convinced that there was really any difference. Suggest what they should do to be more certain that they had got a result that was correct for these ivy leaves. *[2 marks]*

....

....

....

c) The students also did a peel from the upper surface of the leaf but found no stomata on this surface. Suggest how lack of stomata on the upper surface could be an advantage to the ivy plant. *[2 marks]*

....

....

....

(Total = 12 marks)

2   Single-celled algae, such as *Scenedesmus*, can be mixed with a solution of sodium alginate.
    If drops of this mixture are released from a dropper (or a syringe) into a solution of calcium
    chloride, the drops form spherical balls. These are known as algal balls and have a jelly-like
    texture. Thousands of tiny algae are trapped inside each of the balls. The size of the balls can be
    varied by using droppers with different-sized tips.

    The algae contain chlorophyll and can be used to study processes that occur in green plants,
    such as photosynthesis and respiration. They can, for example, be set up in tubes as shown in the
    diagram and used to investigate gas exchange.

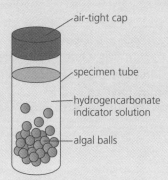

Hydrogencarbonate indicator can
be used to detect changes in carbon
dioxide concentration in the
solution. When the carbon dioxide
concentration is high, the indicator
is yellow. As the concentration of
carbon dioxide decreases, the
colour of the indicator changes to
orange-red, and then to red, and
finally to purple.

    A group of students decided to carry out an investigation using algal balls. They wanted to study
    the effect of different sizes of balls on the rate of gas exchange.

    The students were provided with a suspension of algal balls of four different sizes in separate
    beakers and with a flask of hydrogencarbonate indicator. They had access to standard items
    of laboratory apparatus, such as measuring cylinders, syringes, beakers and plenty of specimen
    tubes. The students also had a card that showed the colour of hydrogencarbonate indicator at
    different concentrations of carbon dioxide.

    a) Describe how you would set up the tubes for the investigation the students want to
       carry out.   *[4 marks]*

    ....

    ....

    ....

    ....

    ....

    ....

    ....

    b) The students left their apparatus all set up overnight, ready to start in the morning. When they
       turned on the lights in the laboratory, they noticed that the indicator was yellow in all the
       tubes. Explain why the indicator was yellow in the tubes.   *[2 marks]*

    ....

    ....

    ....

c) They then kept a light on, close to the tubes, and watched as the colour changed in the tubes.

   i) What colour changes would you expect? Give a reason for your answer.   *[3 marks]*

....

....

....

....

   ii) Describe the measurements you would take to find out whether the size of the algal balls has any effect on the rate of gas exchange taking place in the algal balls. Include reference to any practical details and observations you may have to make.   *[3 marks]*

....

....

....

....

   iii) Suggest the results that you may obtain and give a reason to support your prediction.   *[2 marks]*

....

....

....

(Total = 14 marks)

3 A student did an investigation into transpiration in a plant shoot. She wanted to find out whether the rate of transpiration of a plant shoot was different in the conditions inside a school biology laboratory compared with the conditions outside the laboratory, on the school playing field.

She used a potometer, as shown in the diagram. She recorded the movement of the liquid in the capillary tube and used this as a measure of the rate of transpiration.

In each place, she recorded the air temperature, humidity of the air and wind speed. Her values are given in table 1.

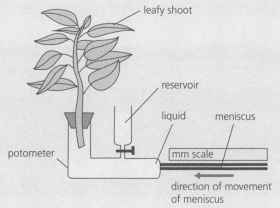

Table 1

Condition	Inside	Outside
Air temperature / °C	23	14
Humidity of air (%)	45	66
Air speed / m per s	0	0.8

In each place, when she set up the potometer, she waited for 5 minutes. She then started to record the position of the meniscus of the liquid in the capillary tube. The readings she took are given in table 2.

She then discussed her investigation and the results she had obtained with other students in the class.

Table 2

Time / minutes	Distance moved by meniscus / mm	
	Inside	Outside
1	3	7
2	6	12
3	10	18
4	13	25
5	17	31

a) She was not sure how to present the results. One student (student A) said it would be best to draw a graph, whereas student B asked why she had taken readings each minute and thought she should just take the distance at 5 minutes as the result.

   i)   What advice would you offer to the student? Give reasons for your answer.   *[2 marks]*

   _____

   _____

   _____

   ii) What calculation could she do to compare the rate of transpiration in each place, using these results?   *[2 marks]*

   _____

   _____

   _____

   iii) What assumption did she make in using this method as a way of measuring the rate of transpiration?   *[1 mark]*

   _____

   _____

b) The students next looked at conclusions that could be drawn from the results in this investigation. They could see that the total distance moved by the meniscus was greater outside, suggesting that the rate of transpiration is faster, but they needed more information to help them draw a suitable conclusion. Here are some of the questions they asked. The letters refer to different students asking the questions.

   C   Why did you leave the potometer for 5 minutes before taking any readings?

   D   How do you know which of the different factors was influencing the rate of transpiration?

   E   Did you use the same leafy shoot each time?

   F   Would differences in light intensity have had an effect on the rate of transpiration in the two places?

   G   Is it enough just to do the investigation with one shoot and take one set of readings?

After this discussion, the students agreed to work together as a group and investigate factors that affect the rate of transpiration. They decided to use the same apparatus but to bring it into the laboratory and make some modifications to the method. They used the questions listed above to help guide them.

i) What could they do to respond to the question from student D?   *[2 marks]*

....

....

....

ii) What could they do to respond to the questions from student E and student G?   *[2 marks]*

....

....

....

....

iii) What answer would you give to student C's question?   *[1 mark]*

....

....

(Total = 10 marks)

● 3.3: Understanding structure, function and processes

1 Water is taken in by the roots of flowering plants. The water then travels up the stem and passes out from the leaves.

a) i) Name the cells in the root that take up water from the surrounding soil.   *[1 mark]*

....

ii) Name the process by which water enters these root cells.   *[1 mark]*

....

iii) What features do these root cells have that adapt them for uptake of water?   *[2 marks]*

....

....

....

b) Water travels up the stem of a plant in the xylem.

i) Describe how features of the cells in the xylem help the water to travel up the stem. *[2 marks]*

....

....

....

ii) What feature of the cells in the xylem helps to provide support for the plant? *[1 mark]*

....

c) i) Name the process by which water passes out from the leaves. *[1 mark]*

....

ii) Name the pores through which water vapour passes out from the leaves. *[1 mark]*

....

iii) These pores can open and close in different conditions. Describe how these changes are brought about. *[3 marks]*

....

....

....

....

....

(Total = 12 marks)

2 a) i) Name the process by which water is lost from the leaf of a plant. *[1 mark]*

....

ii) Explain why a plant takes up less water on a day when it is humid (damp) and the air is still, compared with a day when it is dry and windy. *[4 marks]*

....

....

....

....

....

....

....

b) The diagram shows a section through a leaf of marram grass. This grass grows in sandy places, where the water drains away quickly. It shows adaptations that help it live in dry places. In dry conditions, the leaf rolls up, as shown in the diagram.

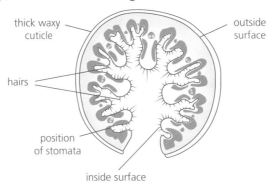

thick waxy cuticle

outside surface

hairs

position of stomata

inside surface

Suggest how features of the leaf help reduce water loss. In your answer, include features of the outside (top) surface and features shown on the inner (lower) side of the leaf.   *[4 marks]*

....

....

....

....

....

....

....

(Total = 9 marks)

3   The diagram shows a simplified view of the human heart and some of the blood vessels in the blood circulatory system. Some of the structures are labelled with the letters A to L. Arrows on the blood vessels show the direction of blood flow.
Use letters on the diagram in your answers to the questions that follow.

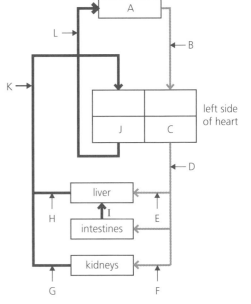

a) Name the structure represented by letter A.   *[1 mark]*

....

b) i)   Name the part of the heart represented by letter C.
[1 mark]

....

ii) Explain why the walls of part J and part C are of different thicknesses.   *[2 marks]*

....

....

....

c) i) Give the letter of the vein that carries oxygenated blood.   *[1 mark]*

    ....

    ii) Explain why most veins carry deoxygenated blood.   *[2 marks]*

    ....

    ....

    ....

d) Suggest why the hepatic portal vein (I) carries high concentrations of glucose and amino acids.

                                                                            [2 marks]

    ....

    ....

    ....

e) i) Give the letter of the blood vessel with the highest blood pressure.   *[1 mark]*

    ....

    ii) Explain why it is important that this blood vessel carries blood at high pressure.   *[1 mark]*

    ....

    ....

f) Explain why there is a difference in the concentration of urea in the blood carried in vessels F and G.   *[2 marks]*

    ....

    ....

    ....

g) i) Give the letter of **one** blood vessel that contains valves.   *[1 mark]*

    ....

    ii) What is the function of valves?   *[1 mark]*

    ....

    ....

h) Explain why high levels of cholesterol in the blood can increase the risk of a person developing coronary heart disease.   *[2 marks]*

    ....

    ....

    ....

                                                                (Total = 17 marks)

● 3.4: Applying principles

1 A scientist was preparing to give a lecture to some students. As the students entered the fairly small room, they saw that a computer trace was being projected on to the wall. The trace showed the concentration of carbon dioxide gas in the air of the room. As more students came in, they watched as the concentration of carbon dioxide continued to rise.

One of the students asked what was happening. The scientist pointed to a sensor lying on the table. This sensor measured the concentration of carbon dioxide in the air in the room. The scientist had previously closed the windows of the room to reduce movement of air between the room and outside.

On the table beside the scientist, he had a large, open, transparent box containing a potted plant. He placed the sensor inside the box, and then sealed the air inside it by placing a lid on the box. He switched on a bright light, which he pointed towards the box. The line on the wall began to fall steadily.

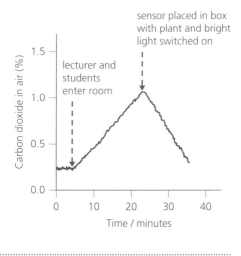

a) i) Explain the rise in carbon dioxide levels that occurred after the students entered the room, as shown on the computer trace. *[3 marks]*

....

....

....

....

....

ii) Explain the fall in carbon dioxide levels inside the box containing the plant when the scientist switched on a bright light. *[3 marks]*

....

....

....

....

....

iii) Suggest what would happen to the levels of carbon dioxide inside the box if the scientist put a light-proof sheet around the box, preventing light from entering it.
Give reasons for your answer. *[2 marks]*

....

....

....

b) The scientist asked the students to suggest what would happen if he took the sensor into a field of maize (corn) so that he could monitor changes in carbon dioxide over a 24-hour period. He told the students to divide the 24 hours into four blocks of 6 hours, as listed:

- 00.00 hrs to 06.00 hrs
- 06.00 hrs to 12.00 hrs
- 12.00 hrs to 18.00 hrs
- 18.00 hrs to 00.00 hrs

i) Sketch a graph to show the changes you would expect for carbon dioxide concentration in the field of maize over a 24-hour period. *[3 marks]*

ii) Give a reason for the changes in carbon dioxide concentration you show on your graph for each of these time periods. *[4 marks]*

....

....

....

....

....

....

....

(Total = 15 marks)

2 A group of students went on a trek in a high mountain area. They travelled by plane to a starting point at an altitude of 3000 m. They then walked for about 10 days, mostly in the region of 4000 m, with some passes at more than 5000 m.

At first they noticed that their breathing rate seemed faster than usual and they felt quite breathless. They didn't walk far before stopping for a rest. After a few days, they felt much better and were able to keep walking at their usual pace. The students realised they were suffering from the effects of lower levels of oxygen in the atmosphere at high altitude.

a) There is about 21% oxygen in the air at all altitudes but, at 4000 m, the reduction in air pressure gives the equivalent of 13% oxygen in the air.

i) Name the gas exchange surface in the lungs. *[1 mark]*

....

ii) Explain how the lower level of oxygen in the air breathed in would affect the rate of oxygen uptake into the blood. *[2 marks]*

....

....

....

....

iii) How does a faster rate of breathing help the students overcome the shortage of oxygen? *[2 marks]*

....

....

....

b) A doctor travelling with the group was able to measure the haemoglobin content of the blood for some students. After three or four days, the students who were tested showed an increase in haemoglobin concentration. This increase was maintained for several weeks after the students had returned home, to lower altitudes.

When they were home again, they did some research and found some typical values for haemoglobin concentration (in males) at different altitudes. These are shown in the table.

Altitude	Haemoglobin concentration / g per dm³
at sea level	148
visitor to Mount Everest at 5790 m	196

i) Calculate the percentage increase in haemoglobin concentration in the visitor to Mount Everest compared with that at sea level. Show your working. *[2 marks]*

Percentage increase = ...

ii) Explain why an increase in haemoglobin concentration helps visitors to high altitudes. *[2 marks]*

....

....

....

....

iii) Suggest why the students were pleased that the data they found in their research agreed with their own measurements made on their mountain trek. *[1 mark]*

....

....

c) Suggest why some athletes undertake training at high altitude in the period leading up to an important athletic event, such as running a marathon. *[2 marks]*

....

....

....

(Total = 12 marks)

4.1: Practical activities

1 Some students investigated responses of cress seedlings to different sources of light.

The students sowed some cress seeds in specimen tubes on damp cotton wool. They then wrapped each tube with a light-proof cover of aluminium foil, except for a small window on one side. They covered this window with transparent plastic film of different colours. This meant the growing seedlings received light of different colours. For tube A, they used a black cover over the window so that the seedlings did not receive any light.

They left the tubes by a window. After four days, the students removed the covers and made observations of the growth of the seedlings. The appearance of the seedlings was as shown in the diagrams of tubes A to E.

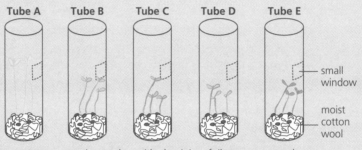

specimen tubes with aluminium foil covers removed

The table below summarises the colours of light for the growing seedlings. The students also used this table to record their results.

Tube	Colour of light	Height of seedlings / cm			Colour of seedlings	Description of angle of growth of seedlings
A	dark (no light)				yellow (less chlorophyll)	
B	white (normal light)				green	growth towards the light (bend from vertical)
C	red				green	
D	green	0.8	0.6	0.6	green	growth upright
E	blue				green	

a) i) Record the heights of the three seedlings in each of the tubes. Write your measurements in the correct spaces in the table. Measurements for the seedlings in tube D have been done for you.   *[4 marks]*

ii) Describe how you made your measurements of height to ensure they are accurate and comparable in all the tubes.   *[2 marks]*

....

....

....

iii) Describe the appearance of the seedlings, giving the direction of growth. Write your descriptions in the correct spaces in the table. Descriptions for seedlings in tubes B and D have been done for you.   *[3 marks]*

b) Seedlings in tube A grew the tallest. Suggest what would happen to the seedlings in this tube if the investigation was continued for several more days. Give a reason for your answer. *[2 marks]*

....

....

....

c) Explain why the seedlings in tube B grew towards the light. *[2 marks]*

....

....

....

d) How could you set up the experiment to make the results more reliable? *[1 mark]*

....

....

(Total = 14 marks)

2 Two students carried out an investigation into their reaction times. They used an electronic clock as a timer. The clock was accurate to 0.01 seconds.

Student A started the clock by using a switch hidden below the table so that student B could not see when it was being started.

As soon as the clock started running, student B had to stop it by pressing a button. The time on the clock showed the reaction time in seconds. They repeated this 10 times.

Here are student B's results, as recorded from the figures read from the clock.

0.11   0.13   0.14   0.12   0.18   0.15   0.13   0.15   0.16   0.15

The students then changed places so that student B started the clock and student A had to stop it. Here are student A's results.

0.12   0.10   0.26   0.16   0.15   0.11   0.14   0.10   0.13   0.16

a) i) Organise the results for both students in a suitable table to present the results. (Use a separate sheet of paper for your table.) *[3 marks]*

   ii) Identify any anomalous results. *[1 mark]*

....

   iii) Calculate the mean value for the reaction time for student B. Show your working. *[2 marks]*

Mean = ...

b) Outline the pathway taken by nerve impulses in the body from the stimulus (seeing that the clock has been started) until the finger presses on the button. Include the names of the neurones in the pathway. *[4 marks]*

....

....

....

....

....

....

....

c) The students decided to use the same method to find out whether listening to music through earphones affected their reaction times. Describe the steps they would take to do this. *[3 marks]*

....

....

....

....

....

(Total = 13 marks)

● 4.2: Understanding structure, function and processes

1 The diagram shows the structure of the human eye.

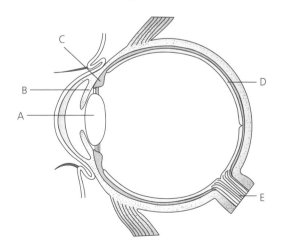

a) The table lists the functions of some parts of the eye. Use letters from the diagram to complete the table, matching the parts of the eye to their functions. Write **one** letter only in each box. *[4 marks]*

Function	Letter of part
controls the amount of light entering the eye	
carries nerve impulses to the brain	
controls the thickness of the lens	
bends (refracts) light rays	
has receptors for light	

b) i) On the diagram, mark the position of the **blind spot** with a label line and the letter **X**. *[1 mark]*

   ii) Explain why an image falling on the blind spot cannot be seen. *[2 marks]*

   ....

   ....

   ....

c) A girl was watching athletics in a large stadium. She looked down at the watch on her wrist. Describe the changes that take place in parts A and C of the eye to bring the image of the watch to a focus on the retina. *[3 marks]*

   ....

   ....

   ....

   ....

   ....

(Total = 10 marks)

2 a) The table shows the names of some hormones, the organs that produce them and one effect of each hormone on the body. Complete the table by filling in the empty boxes. The first row has been done for you. *[8 marks]*

Name of hormone	Organ that secretes the hormone	One effect of the hormone on the body
adrenaline	adrenal gland	increases heart rate
	pancreas	
follicle stimulating hormone (FSH)		
		growth and development of the male sexual organs
progesterone		

b) Hormonal coordination occurs by the transmission of chemical substances through the bloodstream. How is nervous coordination achieved? *[2 marks]*

   ....

   ....

   ....

(Total = 10 marks)

3 a) What is meant by excretion? *[1 mark]*

....

....

b) i) The diagram shows the human urinary system, with parts labelled A, B, C, D and E.

The table lists descriptions of some parts of the urinary system. Match letters from the diagram to their descriptions in the table. Letters may be used more than once.
Write **one** letter only in each box. *[4 marks]*

Description	Letter
stores urine	
contents do not contain urea	
where nephrons are found	
carries oxygen to the kidney	

ii) Name the tube carrying urine away from the bladder. *[1 mark]*

....

c) The hormone adrenaline is produced by the adrenal glands. Draw a label line with an **X** on the diagram to show the position of an adrenal gland. *[1 mark]*

d) Adrenaline is secreted when a person is frightened. Explain **two** ways in which adrenaline prepares the body for immediate action. *[4 marks]*

....

....

....

....

....

....

....

....

(Total = 11 marks)

4   The diagram shows a nephron from a kidney.

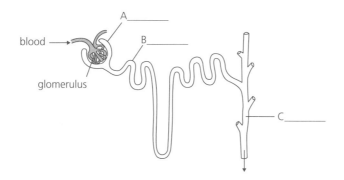

a) i)   Name the parts of the nephron labelled A, B and C on the diagram.   *[3 marks]*

    ii) Describe the process of ultrafiltration, which occurs when the blood passes through the
         glomerulus.   *[2 marks]*

    ....

    ....

    ....

    iii) Give **two** ways in which the glomerular filtrate differs from blood plasma.   *[2 marks]*

    ....

    ....

    ....

    ....

b) Describe what happens to the glucose in the filtrate as it passes through part B.   *[3 marks]*

    ....

    ....

    ....

    ....

c) Describe **two** functions of part C.   *[2 marks]*

    ....

    ....

    ....

    ....

                                                                                    (Total = 12 marks)

4.3: Applying principles

1 Some volunteers took part in an investigation into sweating rates and body temperature when the body was dehydrated.

The volunteers had been deprived of water so that they had become dehydrated to different levels, up to 7% of their normal body water content. They were then exposed to heat and during this time their core body temperatures and their sweating rates were monitored. Temperature of the rectum (rectal temperature) was used as a measure of core body temperature.

The table shows the sweating rates and rectal temperatures of the volunteers at different dehydration levels (0%, 3%, 5% and 7%).

Level of dehydration (%)	Sweating rate / g per m² per hour	Rectal temperature / °C
0	330	37.4
3	300	37.6
5	260	37.8
7	160	38.2

a) Explain how sweating helps to cool the body when a person is hot.   *[3 marks]*

....

....

....

....

....

....

b) i) In this investigation, how does the sweating rate of the volunteers with 7% dehydration differ from the volunteers with 0% dehydration (fully hydrated)?   *[2 marks]*

....

....

....

ii) Suggest an explanation for the difference in sweating rates in the hydrated and dehydrated volunteers.   *[2 marks]*

....

....

....

....

c) What evidence is there in the data provided that the difference in rate of sweating may have affected body temperature? *[2 marks]*

....

....

....

....

d) How would you expect the kidneys of the volunteers to respond to dehydration? *[2 marks]*

....

....

....

(Total = 11 marks)

2 Animals living in deserts need to conserve water. The kangaroo rat is a small mammal that lives in deserts in North America. The rat eats seeds and other dry plant material and never drinks.

A study was made of water balance in a kangaroo rat. The table summarises the water gains and losses of a kangaroo rat during a period of four weeks. During this time, the kangaroo rat consumed 100 g of barley grains for food.

Water gains / cm³		Water losses / cm³	
oxidation water	54.0	urine	13.5
absorbed water (from the barley grains)	6.0	faeces	2.6
		evaporation	43.9
Total	60.0	Total	60.0

a) Water is formed in aerobic respiration from the oxidation of different food molecules, including carbohydrates, lipids (fats) and proteins. Write down a balanced chemical equation to summarise how water is produced from the oxidation of glucose during aerobic respiration. *[3 marks]*

b) The table shows that some water is lost from the kangaroo rat in the urine.

   i) Calculate the percentage of water lost in the urine of the kangaroo rat. Show your working. *[2 marks]*

Percentage water lost = ...

   ii) Where in the kidney is water reabsorbed from the glomerular filtrate back into the blood? *[1 mark]*

....

   iii) The kangaroo rat produces very concentrated urine. Measurements show that the concentration of dissolved substances in the urine of a kangaroo rat is about 3840 arbitrary units, whereas the equivalent figure for the urine of a human is 790 arbitrary units. Suggest how the kidney helps the kangaroo rat survive in the dry conditions in a desert. *[2 marks]*

....

....

....

c) The faeces of a kangaroo rat contained 2.6 g of water per 100 g of barley food consumed. The faeces of a similar rat (which does not live in the desert) contained 13.6 g of water per 100 g of barley food consumed.

   i) Where is water reabsorbed from food material as it passes through the gut in the process of digestion? *[1 mark]*

....

   ii) Use the evidence provided to suggest how the faeces of the kangaroo rat help it to survive in the dry conditions in a desert. *[2 marks]*

....

....

....

d) In the table, the highest figure for water loss is for evaporation. Most of this is in exhaled air from the lungs. Explain why exhaled air contains water vapour and why, as in many mammals, it is important that there is some water vapour in the lungs. *[2 marks]*

....

....

....

(Total = 13 marks)

3 Some people are unable to control their blood glucose level. The condition is known as diabetes and may occur when a person does not produce enough insulin.

A person is sometimes given a glucose tolerance test to help decide whether treatment may be appropriate. The person fasts for 8 to 12 hours, and then the blood glucose level is tested. The person then drinks a large quantity of dissolved glucose and the blood glucose is measured at hourly intervals for several hours.

The graph shows the blood glucose levels of one person with diabetes and one person without diabetes during the 5 hours after a glucose drink.

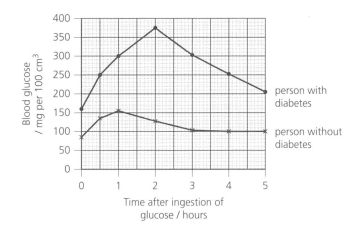

a) i) For the person without diabetes, give values from the graph for each of the following. Write your answers on the short lines. [3 marks]

   1 the normal blood glucose level (at the time the glucose drink is given) ....

   2 the maximum blood glucose level and when this occurs ....

   3 the time taken for the blood glucose level to return to the normal level ....

   ii) Describe how the results for blood glucose level of the person with diabetes differ from the results of the person without diabetes. [3 marks]

   ....

   ....

   ....

   ....

   ....

b) Foods containing carbohydrates may be digested at different rates. If a person has diabetes, suggest why it may be an advantage to eat foods containing carbohydrates (such as starch) that are digested at a slower rate. [2 marks]

   ....

   ....

   ....

(Total = 8 marks)

● 4.4: Extended writing

Write your answers to the extended writing questions on separate sheets of paper. First read the general advice on page 4.

1 An athlete runs a 2000 m race. Describe how heat is produced and lost to keep his body temperature constant during the race. *[6 marks]*

2 Describe the process of ultrafiltration in the kidney nephron. *[6 marks]*

3 Dancers sweat heavily while practising. Describe how antidiuretic hormone (ADH) helps to keep the water content of their blood at normal levels during a rehearsal. *[6 marks]*

5 Reproduction and inheritance

5.1: Using and interpreting data

1 Huntington's disease in humans is an inherited disorder that affects the adult nervous system. It is caused by the dominant allele, **H**. A person without the disease is homozygous for the recessive allele, **h**.

a) Show how the disease may be passed on even if just one parent is heterozygous for Huntington's disease and the other parent shows the normal phenotype for this feature.

Use a genetic diagram to show:

- the genotype of each parent
- the gametes the parents may produce
- the possible genotypes of the children
- the possible phenotypes of the children. *[4 marks]*

b) State the probability of a child of these parents inheriting the condition. *[1 mark]*

....

(Total = 5 marks)

2 The colour of the hairs on the skin of cattle is inherited.

The colour of the hairs is controlled by a gene with two alleles. The alleles are codominant. The allele **R** codes for red hairs and the allele **W** codes for white hairs. Cattle with both alleles in their genotype (**RW**) have a mixture of red and white hairs, and this is called roan.

a) Explain what is meant by the term **codominance**. *[2 marks]*

....

....

....

....

b) A roan cow was mated with a roan bull. Use a genetic diagram to show:

- the genotype of each parent
- the gametes the parents produced
- the possible genotypes of all the offspring
- the possible phenotypes of all the offspring. *[4 marks]*

c) State the probability that one of the offspring was roan. *[1 mark]*

....

d) i) Explain what is meant by the term **heterozygous**. *[2 marks]*

....

....

....

....

ii) Draw a circle around **one** heterozygous individual shown in your genetic diagram. *[1 mark]*

(Total = 10 marks)

3 Diagrams P and Q show the same small portion of the gene that codes for the protein haemoglobin and the steps that take place during protein synthesis. Diagram P shows normal haemoglobin and diagram Q shows a mutant haemoglobin.

Diagram P	Diagram Q
CAT GTA AAC ATA GGA CTT CTT	CAT GTA AAC ATA GGA CAT CTT
step 1 ↓	↓
GUA CAU UUG UAU CCU GAA GAA	X..
step 2 ↓	↓
val his leu thr pro glu glu	Y..

In each diagram, the top row of letters represents base sequences found in the DNA that codes for each haemoglobin molecule. The bottom row of letters in diagram P lists amino acids that correspond to the base sequences above.

Step 1 and step 2 represent events that occur during protein synthesis, between the stages from the base sequence to the assembly of the amino acids in the proteins.

a) i) On diagram P, name steps 1 and 2, which represent stages of protein synthesis. *[2 marks]*

   ii) Name the molecule (in the middle row) that is formed after step 1. *[1 mark]*

   ....

b) In the cell, where are ribosomes found and what is their role? *[2 marks]*

   ....

   ....

   ....

   ....

c) i) Compare the two sequences shown for deoxyribonucleic acid (DNA). On diagram Q, draw a circle around the codon in the **mutant** haemoglobin gene containing a changed base. *[2 marks]*

   ii) Use information in the diagram for synthesis of **normal** haemoglobin to help you complete the following sequences for the **mutant** haemoglobin.

   1 for the molecule in the row labelled **X**

   2 for the amino acid sequence in the row labelled **Y**

   Write your answers on the lines provided. *[4 marks]*

d) In some cases, a change (mutation) in a base within a codon does not lead to a change in the amino acid expected for that codon. Suggest an explanation for this. *[2 marks]*

   ....

   ....

   ....

(Total = 13 marks)

● 5.2: Understanding structure, function and processes

1 The diagram shows the structure of a flower.

a) Name parts B and C. *[2 marks]*

   ....

   ....

b) The table lists some events that occur during reproduction. Complete the table using letters from the diagram to show where each event occurs. Letters may be used more than once. Write **one** letter only in each box. *[4 marks]*

Event	Letter
1 male gametes are produced	
2 a pollen tube begins to grow	
3 fertilisation takes place	
4 the fruit may develop	

c) Give the letters of **two** structures where cell division by meiosis takes place. *[2 marks]*

....

d) i) Explain what is meant by the term **pollination**. *[2 marks]*

....

....

....

ii) Describe **three** ways in which a flower may be adapted for pollination by insects. *[3 marks]*

....

....

....

....

....

....

e) Some plants can reproduce both sexually, by producing seeds, and asexually, by producing runners.

i) Describe **two** ways in which asexual reproduction differs from sexual reproduction. *[2 marks]*

....

....

....

ii) Explain why sexual reproduction can be an advantage to the plant. You can use a diagram in your answer. *[2 marks]*

....

....

....

(Total = 17 marks)

2 The diagram shows the male reproductive organs.

   a) Name parts A and B.   *[2 marks]*

   ....

   ....

   b) Name the tube that carries two different fluids at certain times.   *[1 mark]*

   ....

   c) Draw a label line with an **M** on the diagram to show an organ where meiosis takes place.   *[1 mark]*

   d) i) Name the organ that secretes testosterone.   *[1 mark]*

   ....

      ii) During puberty, testosterone secretion leads to the growth of hair on the body. Describe a
          secondary sexual characteristic, other than hair growth, that also develops during puberty
          in a male.   *[1 mark]*

   ....

   e) i) Explain what is meant by the term **fertilisation**.   *[2 marks]*

   ....

   ....

   ....

      ii) Explain how the X and Y chromosomes, carried by the gametes of the parents, determine the
          sex of a child.   *[3 marks]*

   ....

   ....

   ....

   ....

      iii) When a couple have a child, what is the probability that the child will be a boy?   *[1 mark]*

   ....

(Total = 12 marks)

3 A gardener shows some students how to grow some plants successfully from seed. He explains that
   the soil must be moist, but not so wet that there are no air spaces left in the soil.

   a) i) Explain why a seed needs water to germinate.   *[2 marks]*

   ....

   ....

   ....

   ....

ii) Explain why a seed needs air to germinate. [2 marks]

....

....

....

b) The students prepared some pots of sunflower seeds for germination. They kept one group of pots at 20 °C and another group at 10 °C. Explain why the seeds that were kept at 20 °C germinated before the seeds kept at 10 °C. [2 marks]

....

....

....

c) Name the type of cell division taking place in the root tips of growing seedlings. [1 mark]

....

(Total = 7 marks)

4 The diagram shows the female reproductive organs.
   a) Name parts A and E. [2 marks]

....

....

b) The table lists some events that occur in the female reproductive system.

Event	Letter
1 meiosis takes place	
2 fertilisation takes place	
3 an egg cell is produced	
4 a placenta develops	

Complete the table using letters from the diagram to show where each event occurs. Letters may be used more than once. Write **one** letter only in each box. [4 marks]

c) After the start of puberty, the ovaries secrete oestrogen. Describe **two** secondary sexual characteristics that develop as a result. [2 marks]

....

....

....

....

d) i) Name the organ that secretes progesterone. *[1 mark]*

....

ii) Describe **one** effect of progesterone. *[1 mark]*

....

....

e) The diagram shows changes in the levels of oestrogen and progesterone for the first 14 days of a 28-day menstrual cycle.

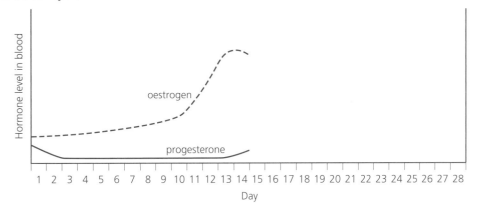

i) Continue the graph lines for oestrogen and progesterone to show the levels that you would expect from day 14 to day 28, if the woman does not become pregnant. *[2 marks]*

ii) The uterus lining becomes thicker during the menstrual cycle. Explain why the thicker lining is important if a fertilised egg cell is to develop into a fetus. *[2 marks]*

....

....

....

(Total = 14 marks)

● 5.3: Applying principles

1 Hair length in cats is controlled by a pair of alleles, and one of the alleles is dominant over the other allele.

A cat breeder keeps genetic records for her cats. The diagram shows part of a family tree showing the distribution of hair length in her cats.

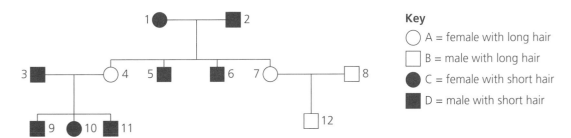

Key

○ A = female with long hair

□ B = male with long hair

● C = female with short hair

■ D = male with short hair

a) What is the hair phenotype of cat 1? *[1 mark]*

....

b) Explain how you can tell that the allele for short hair is dominant. *[2 marks]*

....

....

....

c) What are the hair genotypes of cat 2 and cat 7? Use **H** to represent the allele for short hair and **h** to represent the allele for long hair. *[2 marks]*

....

....

....

d) i) Identify **three** cats that are definitely homozygous for hair length. *[2 marks]*

....

....

ii) Identify **three** cats that are definitely heterozygous for hair length. *[2 marks]*

....

....

....

(Total = 9 marks)

2 Scientists wanted to develop crop plants with resistance to weedkillers. This would allow them to use weedkillers to control weeds but not damage crop plants. The diagram shows one way of doing this.

a) Explain the effects that X-rays may have on the multiplying cells (step 2). *[2 marks]*

....

....

....

b) Cells A, B, C and D came from the same plant and had the same alleles. Suggest why, after treatment with X-rays, plant cell C is able to grow in the dish containing weedkiller (step 4) whereas cells A, B and D are not. *[2 marks]*

....

....

....

c) The scientists were able to grow large numbers of resistant plants from plant C by taking cuttings. Suggest why they chose to take cuttings rather than allowing plant C to produce seeds. *[2 marks]*

....

....

....

(Total = 6 marks)

3 A group of students investigated the change in mass of seeds during and after germination.

They sowed sweet pea seeds in identical pots of soil, putting 15 seeds in each pot. The pots were kept in warm, light conditions and watered regularly. At intervals the students removed 10 seedlings. They were placed in an oven at 100 °C until all water had evaporated from the cells. The students recorded the dry mass of the 10 seedlings.

The results are shown in the following graph.

a) Describe the changes in the dry mass over the 30 days shown in the graph. *[3 marks]*

....

....

....

....

....

....

b) Explain the shape of the graph from day 0 to day 10. *[2 marks]*

....

....

....

c) i) From the graph, on which day did the seedlings begin to carry out photosynthesis? *[1 mark]*

....

ii) Explain your answer. *[2 marks]*

....

....

....

....

(Total = 8 marks)

4 **'Drug resistant tuberculosis has become a major health hazard'**

Headline taken from a report by England's chief medical officer in 2013

Tuberculosis (TB) is a disease in humans caused by bacteria living in the lungs. Antibiotic drugs such as isoniazid affect the metabolism of the bacteria, causing them to die. Occasionally, some TB bacteria mutate and become resistant to one or more drugs and this has led to an increase in the number of human deaths caused by the disease.

a) What is the meaning of the term **mutation**? *[2 marks]*

....

....

....

b) A rare mutation occurs that enables a TB bacterium to survive treatment with an antibiotic. Explain how this may lead to the mutation becoming widespread in the bacterial population. *[4 marks]*

....

....

....

....

....

....

....

....

c) A report on a health service website states, 'Many TB patients fail to complete their antibiotic treatment, a factor that has caused the rise in drug-resistant forms.'

Suggest why completing the course of antibiotics reduces the likelihood of drug-resistant forms multiplying. *[2 marks]*

....

....

....

(Total = 8 marks)

5 Red blood cells are filled with haemoglobin molecules. These are proteins that help to carry oxygen to different parts of the body in the circulatory system.

Sickle cell disease is a genetic disorder and affects red blood cells. It is the result of a mutation in a gene for the protein haemoglobin.

In the red blood cells of affected individuals, the mutant haemoglobin molecules stick together, forming long fibres. As a result, the red cells can become distorted and often appear 'sickle-shaped' under the microscope, giving the disease its name. These distorted red cells can get trapped inside small capillaries, forming blockages. These blockages are painful and they prevent blood reaching the tissues affected.

a) Describe how a mutation in a gene for haemoglobin could lead to the production of sickle cell haemoglobin. *[3 marks]*

....

....

....

....

....

....

b) Suggest why a different mutation in a haemoglobin gene may not have an effect on the phenotype of the individual. *[2 marks]*

....

....

....

c) Name **two** factors that are likely to increase the incidence of mutations.   *[2 marks]*

....

....

....

(Total = 7 marks)

● 5.4: Extended writing

Write your answers to the extended writing questions on separate sheets of paper. First read the general advice on page 4.

1   Describe how pollination and fertilisation take place in a wind-pollinated flower.   *[6 marks]*

2   Describe **two** methods of asexual reproduction in flowering plants and explain the advantages to the plant of asexual reproduction by natural means.   *[6 marks]*

3   Describe the role of hormones in the control of the menstrual cycle.   *[6 marks]*

4   Describe the mechanism by which a gene directs the synthesis of a particular protein within a cell.   *[6 marks]*

5   Describe the theory of natural selection and show how it may lead to evolution.   *[6 marks]*

6 Ecology and the environment

6.1: Using and interpreting data

1 A weather station in Hawaii monitors carbon dioxide concentration in the atmosphere. Data collected there can be used to reflect events occurring on a global scale.

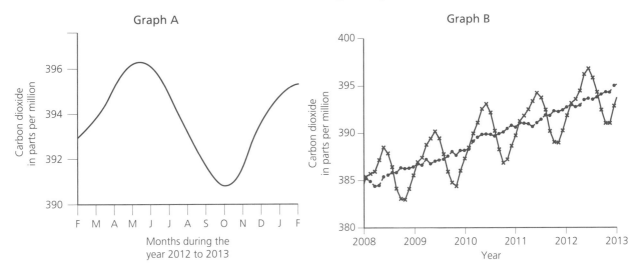

Graph A shows changes in carbon dioxide concentration recorded at this weather station on a monthly basis over a period of one year (February 2012 to February 2013).

a) i)   Give **two** processes that release carbon dioxide into the atmosphere.   [2 marks]

....

....

  ii) Which process in green plants uses carbon dioxide from the atmosphere?   [1 mark]

....

  iii) Suggest why there are fluctuations in carbon dioxide concentration in the atmosphere during the year, as shown in graph A.   [3 marks]

....

....

....

....

....

....

b) Graph B shows data for carbon dioxide concentration recorded at this weather station over a period of five years, from 2008 to 2013. The line with dots (•) shows the average of the fluctuations during each year.

Carbon dioxide is described as a greenhouse gas and contributes to the greenhouse effect.

i) Explain how greenhouse gases contribute to global warming. *[3 marks]*

....

....

....

....

....

....

ii) Name **two** greenhouse gases, other than carbon dioxide and water vapour, and give **one** source of each of the gases you name. *[4 marks]*

....

....

....

....

c) Graph B indicates a steady increase in carbon dioxide over the five years shown. Scientists have evidence for a similar trend of increasing carbon dioxide concentration over the past 150 years.

Describe how human activities may have contributed to this increase in global carbon dioxide concentration. *[3 marks]*

....

....

....

(Total = 16 marks)

2 The table lists a selection of animals and information about some of the foods they eat. All the animals are found in woods in Europe. Living (green) plants are also present in the woods. Dead plant material, such as dead leaves, is found on the ground in the woods.

Animal	Food the animal may eat
ants	caterpillars, earthworms, earwigs
blue tits	ants, aphids, caterpillars, earwigs
caterpillars	living plants
earthworms	dead plant material
earwigs	dead plant material, living plants
ground beetles	ants, caterpillars, earthworms
robins	ants, caterpillars, earthworms
sparrowhawks	blue tits, robins
tawny owls	earthworms, ground beetles, woodmice
woodmice	nuts, berries, caterpillars, earthworms, earwigs

Use information in the table to answer the questions below.

a) i) Write down **one** food chain with four organisms, including a living plant and a sparrowhawk. Use arrows to show the direction of energy flow. *[3 marks]*

ii) What is the source of energy for organisms in this food chain? *[1 mark]*

....

b) i) Write down **one** food chain with at least **three** (living) organisms, including an earthworm (a detritivore) and a tawny owl. *[1 mark]*

ii) Use this food chain to explain how detritivores help to introduce an energy source for organisms in the food web. *[2 marks]*

....

....

....

c) Explain why feeding relationships are often summarised in a food web rather than a food chain. Use examples in the table to help in your explanation. *[3 marks]*

....

....

....

d) A pyramid of energy represents the energy contained in each trophic level. Explain why a pyramid of energy for a food web always has a wide base and a narrow top. *[3 marks]*

....

....

....

....

....

(Total = 13 marks)

6.2: Practical activities

1 Daisy plants have small white flowers and they are easily seen and recognised in areas of short grass.

Some students noticed that daisies were growing in two different areas of grass in their school grounds. Students often walked across one of the areas (area A) on their way to different buildings in the school so they had trampled on the grass in this area. The other area (area B) was out of the way and people rarely walked there.

The students decided to compare the effect of trampling on the distribution of daises in these two areas. They used a quadrat (0.5 m × 0.5 m) to sample each area. They counted the number of daisies in the quadrats. They recorded the number of daisies in 10 quadrats in each of the areas. Their results are shown in the table.

Quadrat number	Total number of daisy flowers in each quadrat	
	Area A (trampled)	Area B (not trampled)
1	0	7
2	15	4
3	8	11
4	12	7
5	5	3
6	10	6
7	9	5
8	2	3
9	16	8
10	9	9
Total		**63**

a) The total number of daisies counted in area B (the untrampled area) is given in the table.

i) Calculate the total number of daisies in area A (the trampled area). Write your answer in the table.

[1 mark]

ii) The density of daisy flowers in area A is 34.4 per m$_2$. Calculate the density of daisy flowers in area B.

[2 marks]

Density of daisy flowers = ...

b) i) How has trampling affected the number of daisy plants in these grass areas? *[1 mark]*

....

ii) Suggest why trampling may have influenced the number of daisy flowers in the two areas.

[2 marks]

....

....

c) When doing the investigation, the students used a random method for placing the quadrats.

i) Explain why the students used quadrats in their investigation. [2 marks]

....

....

....

ii) Why is it important to use a random method for placing the quadrats? [1 mark]

....

....

iii) Describe how the quadrats could be placed randomly. [2 marks]

....

....

....

....

d) i) Area A (trampled grass) shows more variation than area B (untrampled grass) in the number of daisy flowers per quadrat. Suggest a reason for this. [1 mark]

....

....

ii) Suggest **two** factors other than trampling that may have influenced the presence of daisy flowers in each area. [2 marks]

....

....

....

(Total = 14 marks)

2 Some students investigated the growth of organisms on the trunks of trees growing in a wood in a damp area. They recognised four types of organisms growing on the tree trunks: algae, foliose (leafy) lichens, crustose (flat) lichens and moss.

The students compared the growth of these organisms on two species of tree growing in the wood: ash and sycamore. They used quadrats with a grid inside the quadrat and made an estimate of percentage cover of these organisms on the trunks of several trees. If there was nothing growing, they recorded bare bark. The quadrat on a tree trunk is shown in the diagram.

Their combined results from several trees are shown in the pie charts.

Percentage cover of species on ash tree bark

Percentage cover of species on sycamore tree bark

Key
- bare bark
- algae
- moss
- crustose lichen
- foliose lichen

a) i) When collecting their data, suggest why the students used several trees of each species. *[1 mark]*

....

....

ii) Suggest the advantage of using quadrats with a grid inside. *[1 mark]*

....

....

iii) Give **one** variable the students should control when using the quadrats and give a reason for your answer. *[2 marks]*

....

....

....

....

b) Two students decided to present the results in a bar chart rather than in a pie chart.

i) Draw a bar chart so that you can compare percentage cover of each category of organism recorded on these trees. (Use a separate sheet of graph paper to plot your bar chart.) *[5 marks]*

ii) From this investigation, which species of tree showed greatest variation in cover of the different organisms growing on the trunk? Use figures from the results to support your answer. *[2 marks]*

....

....

....

c) The students knew that algae are protoctists and learnt that lichens are an association between an alga and a fungus growing together. Mosses are simple plants that lack xylem and grow only to a small size.

i) Which of the organisms listed above contains no chlorophyll? Suggest how this organism obtains its food supply. *[2 marks]*

....

....

....

ii) Suggest why moss plants grow only to a small size. *[2 marks]*

....

....

....

(Total = 15 marks)

3 *Pleurococcus* is an alga (a protoctist) that often grows on the bark of trees. It forms a film and when it is dry it looks like a green powder. *Pleurococcus* is single-celled, contains chlorophyll and requires light and moisture to carry out photosynthesis.

Some students investigated the abundance of *Pleurococcus* on different sides of the trunk of a tree. They lived in an area where the prevailing winds carrying rain blow from the west. They predicted that the alga would grow best on the west side of a tree trunk and less well on the north and south sides.

They chose three ash trees (A, B and C) and tied a string around each tree at a height of 120 cm. On three sides of each tree (north, south and west), they marked an area 10 cm × 10 cm just below the string. They scraped the *Pleurococcus* off the bark from inside each of these areas into pieces of folded paper. This is shown in the diagram.

To estimate the quantity of *Pleurococcus* in each area, they transferred the powder collected in each paper into a boiling

tool being used to scrape algae from tree trunk

east   south

north

string

10 cm × 10 cm square

sheet of paper collecting scraped algae from tree trunk

tube containing 10 cm³ of ethanol (alcohol). They placed the boiling tubes in a boiling water bath for 2 minutes. The ethanol in each tube became green as the chlorophyll pigments were extracted from the algae. The intensity of the green colour in the ethanol was matched with a colour chart and scored from 0 (colourless) to 10 (very green).

The table shows their results.

Direction faced by side of tree	Colour score of green extract		
	Tree A	Tree B	Tree C
north	2	0	3
south	1	0	2
west	5	4	5

a) i) How many boiling tubes of ethanol with *Pleurococcus* samples did they place in the boiling water bath? [1 mark]

....

ii) Describe **one** safety precaution they should have taken when carrying out this investigation and explain why the precaution should be taken. [2 marks]

....

....

....

b) i) How far do their results support their predictions? Use figures in the table to support your answer. [3 marks]

....

....

....

....

....

ii) Suggest why they predicted more algal growth on the west side of the trees. [3 marks]

....

....

....

....

....

iii) How could they make their results more reliable? [1 mark]

....

(Total = 10 marks)

6.3: Applying principles

1 A student investigated the effects of plant species on the biodiversity of animals in leaf litter (animals associated with the fallen leaves on the soil under trees) in a woodland.

She chose two areas (J and K) where different species of trees were present and these determined the types of tree leaves that had fallen on the floor of the woodland:

Area J – only oak trees

Area K – a mixture of ash, oak, hawthorn and hazel

In each area she measured the light level with a light meter and the soil temperature with a soil thermometer probe.

The student collected a sample of 100 leaves from the ground in each study area. She counted the number of leaves from each species of tree in each sample. She used this as a measure of the biodiversity of tree leaves in each area.

The student collected animals using pitfall traps, as shown in the diagram. These were buried in each area and left overnight. The following day, she emptied each pitfall trap into a white tray and then classified the animals into the groups (A to D) listed below:

A – **Plant sap feeders**: aphids; plant bugs
B – **Plant leaf feeders**: snails; slugs; caterpillars
C – **Carnivores**: centipedes; ants; earwigs; spiders
D – **Detritivores**: woodlice; beetles

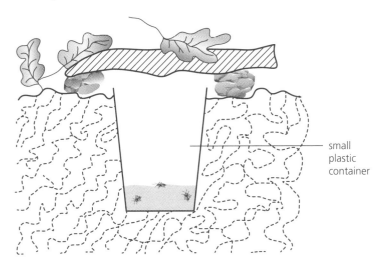

a) Suggest how the student should have made the light measurements to provide reliable data.

[2 marks]

....

....

....

....

b) Suggest how the student may have collected the leaves from each area to ensure the data were representative of each area. [3 marks]

....

....

....

....

....

c) i) Describe how a pitfall trap is set up and suggest why the trap should be emptied after 24 hours. [4 marks]

....

....

....

....

....

....

....

ii) Suggest why no flying insects were found in the traps. [1 mark]

....

....

d) In both areas, one group of animals was always much rarer than the others. Which group (A to D) would you predict to be the rarest? Give a reason for your choice. [2 marks]

....

....

....

e) The light levels near the woodland floor were much higher in area K. Suggest how this abiotic factor difference may have affected the temperature of the leaf litter. [2 marks]

....

....

....

f) Predict which area (J or K) would have the greatest animal biodiversity, giving reasons to support your answer. *[4 marks]*

....

....

....

....

....

....

....

(Total = 18 marks)

● 6.4: Extended writing

Write your answers to the extended writing questions on separate sheets of paper. First read the general advice on page 4.

1 A pyramid of energy is a way of representing the total energy contained in the organisms in a community. Explain why a pyramid of energy is wide at the base but becomes progressively narrower with each trophic level towards the top. *[6 marks]*

2 Describe the part played by microorganisms in the carbon cycle and in the nitrogen cycle. *[6 marks]*

3 A person running across grassland uses the process of respiration to release energy from glucose in the leg muscles. Describe how a carbon atom in a glucose molecule in the runner's leg can become part of a muscle in a rabbit, an animal that is a primary consumer and lives in the surrounding grassland. *[6 marks]*

Use of biological resources

7.1: Using and interpreting data

1. When fruits such as apples or bananas are harvested, certain living processes continue, even though the fruit has been removed from the plant.

   Commercial fruit growers store the harvested fruit for several weeks (or months) before it is sold in shops and markets. During this time, as the fruits ripen, they may lose weight, and change in colour, texture and flavour. The grower aims to ensure that the stored fruits remain in good condition for sale and to minimise losses from fruit that has gone bad (spoiled).

   Inside the storage containers, humidity, temperature and the composition of the atmosphere can often be controlled. Different ways of packaging fruits also help to keep them in a suitable condition.

   An investigation was carried out into the effect of different wrapping and packaging materials on plantains (a type of banana). They were stored at tropical temperatures (about 30 °C) and in a relative humidity between 60% and 80%. The results of the investigation are summarised in the table.

Packaging materials	Time to ripeness / days	Loss in mass at ripeness compared with mass at harvest (%)
not wrapped	15.8	17.0
paper	18.9	17.9
perforated polythene	26.5	7.2
polythene	36.1	2.6

   Note: Perforated polythene has many tiny holes.

   a) i) Plot these results as a bar chart on a graph grid. Use one vertical axis for *Time to ripeness / days* and the other vertical axis for *Loss in mass at ripeness (%)*. (Use a separate sheet of graph paper to plot your graph.) *[6 marks]*

      ii) Describe the effects of different packaging on the length of time it takes for the plantains to ripen. Use examples in the table to support your answer. *[2 marks]*

      ....

      ....

      ....

   b) i) Name **two** processes that would continue in the plantains after harvest and are likely to lead to a loss in mass. In each case, name the substance(s) that would be lost from the plantains. *[2 marks]*

      ....

      ....

      ....

      ii) Which type of packaging was least successful in reducing the loss of mass? Suggest why this packaging was not effective. *[2 marks]*

      ....

      ....

....

....

c) Suggest **two** reasons why fruit may become spoiled during storage so that it is unsuitable for sale. *[2 marks]*

....

....

....

d) On the basis of this investigation, what advice would you give to a fruit grower who produced fruit on a small scale and wanted to keep the fruit in good condition as long as possible? *[1 mark]*

....

....

(Total = 15 marks)

2 *Lactobacillus* bacteria are involved in a number of fermentations. Yoghurt is made by the fermentation of milk and *Lactobacillus* bacteria are involved in the fermentation process.

Silage is made by the fermentation of grass. Different *Lactobacillus* bacteria are involved in the fermentation process. Making silage preserves grass so that animals (such as cattle) can feed on the silage a year or more after the grass is cut.

In both fermentations, lactic acid is produced from sugars present in the milk (in making yoghurt) and in the grass (in making silage). The quality of the silage is better if the fermentation takes place quickly.

When making silage, farmers often add a mixture containing bacteria, including *Lactobacillus*, to help speed up the fermentation process. A comparison was made of the effect of adding the mixture with *Lactobacillus* to a sample of grass. Sample A was the control grass sample, without added *Lactobacillus*, and sample B was the grass sample, with added *Lactobacillus*.

Graph 1 shows pH changes over a period of 30 days of grass being converted to silage with and without the addition of the bacterial mixture. Graph 2 shows the production of lactic acid in the same fermentation.

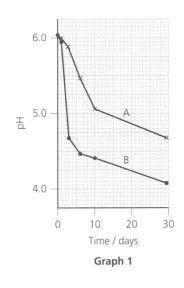

Graph 1

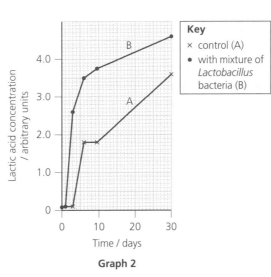

Graph 2

a) i) In graph 1, at 6 days, in grass sample A (without added *Lactobacillus*) the pH has fallen from 6.0 to 5.5. What is the pH in grass sample B (with added *Lactobacillus*) at 6 days?   [1 mark]

....

ii) In graph 2, calculate the difference in lactic acid production in grass A (without added *Lactobacillus*) and grass B (with added *Lactobacillus*) at 6 days. Show your working.   [3 marks]

....

....

....

....

....

....

b) i) Give evidence from the graph to support the view that the fall in pH during the fermentation is the result of the production of lactic acid.   [2 marks]

....

....

....

ii) In the fermentation of milk for production of yoghurt, how does the fall in pH affect how thick the milk is?   [1 mark]

....

c) Making silage is a way of preserving cut grass over a long period of time as food for cattle. Give **two** reasons why yoghurt is produced as a food for humans.   [2 marks]

....

....

....

d) i) Why does some fermentation occur in grass sample A, when no *Lactobacillus* bacteria are added?   [1 mark]

....

....

ii) Suggest why farmers often cut grass for silage on a sunny day and in the afternoon.   [2 marks]

....

....

....

....

(Total = 12 marks)

3 The term **aquaculture** describes the farming of aquatic organisms, including fish, molluscs (such as snails) and crustaceans (such as prawns and shrimps). Aquatic organisms are farmed in fresh water and in marine waters. Most farmed fish is consumed by people, although some is converted into feed for animals or used in other ways.

The table shows the world aquaculture production for 10 years, from 2001 to 2010. Figures show mass of farmed aquatic organisms in millions of tonnes.

Year	Mass in millions of tonnes		
	Freshwater	Marine	Total
2001	21.8	12.8	34.6
2002	23.3	13.5	36.7
2003	24.9	14.0	38.9
2004	27.2	14.7	41.9
2005	29.1	15.2	44.3
2006	31.3	16.0	47.3
2007	33.4	16.6	49.9
2008	36.0	16.9	52.9
2009	38.1	17.6	55.7
2010	41.7	18.1	59.9

a) Describe the changes shown in total aquaculture production over the 10 years from 2001 to 2010. *[1 mark]*

....

....

b) i) On a graph grid, plot a line graph to show the data for both freshwater and marine aquaculture production over the years 2001 to 2010. Plot both sets of data on the same grid. Use a ruler to join the points with straight lines. (Use a separate sheet of graph paper to plot your graph.) *[6 marks]*

ii) In 2001, aquaculture production in fresh water was 63% of the total production. Calculate the percentage production in fresh water in 2010. Show your working. *[2 marks]*

Percentage production in freshwater = ...

iii) Suggest reasons for any changes shown by these graphs over the years 2001 to 2010. *[2 marks]*

....

....

....

c) i) Suggest **two** advantages of producing fish by aquaculture rather than catching wild fish. *[2 marks]*

....

....

....

ii) Describe **two** ways in which aquaculture may harm the environment. *[2 marks]*

....

....

....

....

(Total = 15 marks)

7.2: Practical activities

1 A grower in England had a farm where she grew vegetables. She grew some inside a polytunnel and some outside in a field. She sold the vegetables direct to local shops and markets.

The figures in the list show the yields she obtained with French beans in 2011. Inside the polytunnels, the beans were grown in rows, with a total length of 40 m. Outside in the field, they were grown in rows, with a total length of 100 m. She picked the beans when they were ready and weighed them to find the mass in kg each time she harvested them.

a) i) Draw a suitable table that you can use to present the data so that you can compare the total yield of the beans inside the polytunnel and outside in the field. Then organise the data in your table. (Use a separate piece of paper to draw your table.) *[4 marks]*

ii) Calculate the total yield of beans harvested from each place on the farm. Show your working. *[2 marks]*

Yield of beans = ...

Mass of beans harvested in kg (dates of harvest are given in brackets)

June (inside)

0.85 (10th); 0.75 (11th); 0.8 (13th); 1.6 (17th); 4.6 (19th); 1.2 (21st); 2.1 (23rd); 2.2 (24th); 4.8 (27th); 3.1 (28th); 1.4 (29th)

July (inside)

5.3 (3rd); 5.8 (6th); 2.0 (11th); 1.6 (13th); 1.0 (14th); 1.6 (16th); 0.75 (31st)

July (outside)

3.5 (10th); 2.0 (11th); 1.6 (21st); 2.0 (23rd); 1.0 (27th); 0.6 (31st)

August (inside)

2.2 (5th); 0.75 (7th); 2.0 (11th); 1.5 (14th); 1.8 (17th); 1.8 (20th); 2.0 (23rd); 0.8 (28th)

August (outside)

3.2 (3rd); 2.4 (5th); 1.5 (7th); 4.5 (9th); 4.4 (14th); 1.1 (17th); 1.0 (25th); 0.5 (28th)

September (outside)

0.8 (5th); 0.6 (8th); 0.6 (14th)

data courtesy of Longs Farm, Hartest.

b) i) The total yield of beans from inside the polytunnel is 1.36 kg per m of row. Calculate the total yield of beans grown outside in the field in kg per m of row. *[1 mark]*

....

ii) Suggest reasons for the difference in yield of beans in the polytunnel compared with the field. *[2 marks]*

....

....

....

....

c) The grower sowed the bean seeds in trays then planted out the young plants when they had germinated and grown to a suitable size. The table shows the dates on which she sowed the seeds and when she planted them out.

Place on farm	Sowed bean seeds	Planted out seedlings	Harvest period	
			Start	Finish
inside polytunnel	22 March 8 April	16 to 20 April		
outside in field	18 April 16 May	11 May 14 May 11 June		

In the spaces in the table, write in the dates for the start and finish of the harvest periods for the beans grown in both places. *[2 marks]*

d) Use the information provided as well as your own calculations to answer these questions.

i) Suggest **two** advantages for the grower of growing the beans in the polytunnel. *[2 marks]*

....

....

....

....

ii) Suggest **one** way that the grower benefits from growing the beans both inside the polytunnel and outside in the field. *[1 mark]*

....

....

(Total = 14 marks)

2 The steps shown in the diagram summarise a procedure that can be used in a school or college laboratory to clone cauliflowers (a vegetable). It illustrates the essential stages of micropropagation techniques used in commercial laboratories, where the procedure is carried out under carefully controlled sterile conditions.

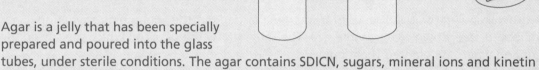

SDICN is a sterilising solution, similar to that used for sterilising babies' bottles and for small amounts of drinking water.

Agar is a jelly that has been specially prepared and poured into the glass tubes, under sterile conditions. The agar contains SDICN, sugars, mineral ions and kinetin (a plant growth regulator).

Step 1: Clean bench with alcohol and paper towel.
Step 2: Dip forceps and scalpel (knife) in beaker containing sterilising solution.
Step 3: Use scalpel to cut a small 'floret' from the head (white part) of a cauliflower and place it on the lid of a sterile Petri dish.
Step 4: Cut this floret lengthwise into several smaller pieces, about 3 mm to 5 mm long.
Step 5: Use forceps and transfer these smaller pieces into a beaker containing SDICN. Leave them in the SDICN for 15 minutes. Shake the beaker gently for 5 seconds in each minute.
Step 6: Use forceps and transfer the small pieces into separate tubes containing the agar. Press the stalk into the agar. Replace the lid on each tube.
Step 7: Place the tubes in a warm place (say, about 35 °C) and in the light.
Finally: After about 10 days the small pieces should begin to look green and show some growth. Later, roots and shoots begin to develop.

a) i) List **three** steps when the sterilising solution (SDICN) is used. *[1 mark]*

....

ii) Explain why it is necessary to use sterile conditions. *[2 marks]*

....

....

....

b) i) In micropropagation techniques, what name is given to the 'smaller pieces' cut from the cauliflower? *[1 mark]*

....

ii) Why are sugars included in the agar in the tubes? *[1 mark]*

....

iii) Explain why the tubes are left in the light. [2 marks]

....

....

....

c) The method described is similar to one used by research organisations interested in the conservation of rare and endangered plants. A scientist working in remote places can take the equipment into the field. Without damaging the plant, the scientist can follow the procedure and remove small parts of the endangered plants, getting them into the tubes as described for step 6. The scientists then take the tubes back to their laboratory to allow them to grow into whole plants.

i) Suggest why scientists wish to conserve rare and endangered species of plants. [2 marks]

....

....

....

ii) Explain why micropropagation techniques are useful in this conservation work. [2 marks]

....

....

....

iii) Suggest the advantages for the scientists of using the method and equipment described and collecting their small pieces of plants for micropropagation in the field. [2 marks]

....

....

....

(Total = 13 marks)

3 Some students carried out an investigation into the effect of fertiliser on the growth of grass seedlings.

They grew the grass in small compartments in a seed tray, as shown in the following diagram. Each of the trays had 10 compartments and they gave the same treatment to all the compartments of a tray. Each day the students checked that there was enough water in the tray and added more if needed. The sand contains no nutrients.

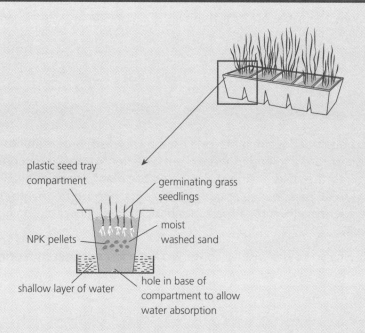

plastic seed tray compartment

germinating grass seedlings

NPK pellets

moist washed sand

shallow layer of water

hole in base of compartment to allow water absorption

The students added the fertiliser as NPK pellets. NPK indicates that the pellets contain nitrogen, phosphorus and potassium. The pellets release the fertiliser slowly, so that the supply is continuous throughout the investigation. The number of pellets added for each tray is shown in the table. They added 10 grass seeds to each compartment.

After two weeks, they measured the height of each individual grass seedling in all the compartments. They measured the height from the top of the sand in the seed tray to the highest leaf tip of the grass. They calculated the mean height for each treatment (number of pellets). Their results are shown in the table.

Number of NPK pellets in each compartment	Mean height of seedlings / mm
0	75
5	165
10	180
20	156
30	60

a) i) In total, how many grass seeds did the students use in this investigation? Show how you work out your answer. *[2 marks]*

ii) Suggest why the students decided to use this number of seeds. *[1 mark]*

....

....

iii) Suggest why the students calculated the mean of their individual measurements. *[2 marks]*

....

....

....

b) Plot a line graph on a grid of graph paper 10 cm high by 8 cm wide to show how fertiliser affected the growth of the grass seedlings in this investigation. Use a ruler to join the points with straight lines. (Use a separate sheet of graph paper to plot your graph.)　　　*[5 marks]*

c) i)　Give **two** factors that the students should have kept constant for all seedlings when doing this investigation.　　　*[2 marks]*

....

....

....

....

　　ii)　Give **two** possible sources of error in the way that the students carried out this investigation.　　　*[2 marks]*

....

....

....

....

d) Using the results of this experiment, what advice would you give to a farmer wishing to use fertiliser on the crops being grown on a farm?　　　*[2 marks]*

....

....

....

....

(Total = 16 marks)

7.3: Understanding structure, function and processes

1 When tomatoes ripen, they turn red, become sweeter and develop other flavours, making them good to eat. At the same time, material between the cell walls begins to go soft. An enzyme called polygalacturonase (PG) controls this reaction.

Growers often pick their tomatoes before they are ripe because they have not yet started to go soft. This means that they are less likely to get damaged when they are transported to shops and markets where they are sold. However, often the flavour of tomatoes is not so good before they are fully ripe.

Scientists have produced a genetically modified (GM) tomato in which the PG enzyme is much less active than in the normal tomato. This means that the growers pick the tomatoes when they are ripe and taste better, but they do not go soft so quickly. This reduces losses after harvest from spoiled tomato fruits.

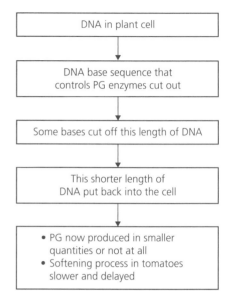

The flow chart summarises the steps taken to produce the GM tomatoes.

a) i) Where is DNA found in a cell?   *[1 mark]*

....

ii) Name **two** bases that might be in the sequence cut out of the DNA.   *[1 mark]*

....

b) i) Name the type of enzyme that is used to cut DNA in GM techniques.   *[1 mark]*

....

ii) Suggest how the shortened length of DNA might be put back into the tomato cell.   *[2 marks]*

....

....

....

c) i) Suggest why altering the PG enzyme affects the softening of the tomatoes.   *[1 mark]*

....

ii) Explain why these GM tomatoes are not 'transgenic'.   *[2 marks]*

....

....

....

d) i) Give **two** advantages to the grower of growing GM tomatoes. *[2 marks]*

....

....

....

ii) These GM tomatoes were banned from being sold in certain countries because people objected to the use of GM crops. Suggest one reason why some people object to GM crops. *[1 mark]*

....

....

(Total = 11 marks)

2 Scientists have developed techniques that they can use to clone mammals. Dolly (a sheep) was an early example of a successfully cloned mammal. Dolly was born in 1996. Several other mammals have now been cloned, including a dog (Snuppy), a cat (Copy Cat), a gaur (an endangered species, related to cattle), a camel and a goat.

The statements A to G describe stages in the technique used for cloning, but the statements are not in the correct sequence.

Statements

A Skin cells are taken from the original animal to be cloned.

B The embryo develops into a foetus and is born.

C The embryo grows on an artificial medium.

D The egg cell containing the skin cell nucleus develops into an embryo.

E The nucleus is removed from an egg cell so that this becomes an enucleated cell.

F The embryo is implanted into a surrogate mother.

G The skin cell nucleus is inserted into the egg cell.

a) Arrange these stages in the correct sequence and write your answers in a table For each stage, state whether the cell(s) are haploid, diploid or have no nucleus (no chromosomes). Stage A is the first, but the rest are out of order. *[5 marks]*

b) Give **two** ways in which sexual reproduction in mammals differs from the processes used for production of offspring in cloning techniques used with mammals.   *[2 marks]*

....

....

....

....

c) Suggest **two** reasons why cloning of mammals may be undertaken.   *[2 marks]*

....

....

....

....

(Total = 9 marks)

● 7.4: Extended writing

Write your answers to the extended writing questions on separate sheets of paper. First read the general advice on page 4.

1   Explain how glasshouses and polytunnels can be used to grow crops, such as tomatoes or peppers, in a way that the grower benefits from increased yields.   *[6 marks]*

2   Explain how microorganisms are beneficial to the lives of people. Include suitable examples to illustrate your answer.   *[6 marks]*

3   Explain how artificial selection differs from natural selection as a way of changing the characteristics of populations of plants and animals. Include examples to illustrate your answer.   *[6 marks]*

8 Experimental design (CORMS)

Write your answers to these CORMS questions on separate sheets of paper. First read the general advice on page 5.

1 When bananas ripen, their skin turns from green (for unripe bananas) through a series of colours to yellow (for ripe bananas). For fruits, a gas called ethene helps with the ripening process. Ripe fruits give off ethene.

  Design an investigation to find out whether placing a ripe tomato with green bananas speeds up the ripening of the bananas.   *[6 marks]*

2 A group of farmers grew cabbages in their fields. They wanted to increase the yield of their crop and one farmer suggested they use fertiliser.

  Design an investigation to find out whether using animal manure as the fertiliser increases the yield of the cabbages.   *[6 marks]*

3 When yoghurt is made, fruit is sometimes added. This fruit can be included at the start of the fermentation or after incubation is complete.

  Design an investigation to find out whether adding fruit (such as peaches) at the start of the process affects the time it takes for the yoghurt to set.   *[6 marks]*

4 A group of students stood in pairs in a circle on a games field. A teacher in the middle blew a whistle. When the whistle blew, one student in the pair started to jump up and down on the spot. The other student in the pair had a stopwatch and noted how quickly the partner responded to the sound of the whistle.

  Design an investigation to find out whether reaction times are different if the student is blindfolded (e.g., wearing a dark bag over their head).   *[6 marks]*

Notes

INTERNATIONAL GCSE (9–1)

Biology

for Edexcel International GCSE

Maximise your performance with practice questions, written to support and enhance the content of the Edexcel International GCSE (Subject) book.

◆ Enhance learning with extra practice designed to support the Student Book.
◆ Test knowledge with a variety of exam-style questions including multiple choice and extended writing.
◆ Perfect for homework and independent study to ensure you have understood concepts.

www.hoddereducation.co.uk

ISBN 978-1-5104-0565-3

9 781510 405653